229
Topics in Current Chemistry

Springer
Berlin
Heidelberg
New York
Hong Kong
London
Milan
Paris
Tokyo

New Aspects in Phosphorus Chemistry III

Volume Editor: Jean-Pierre Majoral

With contributions by
T. Chivers, K.V.L. Crépy, H.-J. Cristau, M. Gulea,
T. Imamoto, K. V. Katti, K. Lammertsma, S. Masson,
N. Pillarsetty, K. Raghuraman, W. W. Schoeller,
M. Taillefer

Springer

The series *Topics in Current Chemistry* presents critical reviews of the present and future trends in modern chemical research. The scope of coverage includes all areas of chemical science including the interfaces with related disciplines such as biology, medicine and materials science. The goal of each thematic volume is to give the non-specialist reader, whether at the university or in industry, a comprehensive overview of an area where new insights are emerging that are of interest to a larger scientific audience.

As a rule, contributions are specially commissioned. The editors and publishers will, however, always be pleased to receive suggestions and supplementary information. Papers are accepted for *Topics in Current Chemistry* in English.

In references *Topics in Current Chemistry* is abbreviated Top. Curr. Chem. and is cited as a journal.

Springer WWW home page: http://www.springer.de
Visit the TCC home page at http://www.springerlink.com/series/tcc/

ISSN 0340-1022
ISBN 3-540-00714-8
DOI 10.1007/b11214
Springer-Verlag Berlin Heidelberg New York

Library of Congress Catalog Card Number 74-644622

Springer-Verlag Berlin Heidelberg New York
a member of BertelsmannSpringer Science+Business Media GmbH

http://www.springer.de

Cover design: KünkelLopka, Heidelberg/design & production GmbH, Heidelberg
Typesetting: Fotosatz-Service Köhler GmbH, 97084 Würzburg

02/3020 ra – 5 4 3 2 1 0 – Printed on acid-free paper

Volume Editor

Dr. Jean-Pierre Majoral
Directeur de Recherche
Laboratoire de Chimie de Coordination du CNRS
205, route de Narbonne
31077 Toulouse, Cedex 4, France
E-mail: majoral@lcc-toulouse.fr

Editorial Board

Topics in Current Chemistry Also Available Electronically

For all customers with a standing order for Topics in Current Chemistry we offer the electronic form via SpringerLink free of charge. Please contact your librarian who can receive a password for free access to the full articles by registration at:

http://www.springerlink.com/orders/index.htm

If you do not have a standing order you can nevertheless browse through the table of contents of the volumes and the abstracts of each article at:

http://www.springerlink.com/series/tcc/

There you will also find information about the

- Editorial Board
- Aims and Scope
- Instructions for Authors

Preface

The present volume on New Aspects in Phosphorus Chemistry is the third one devoted to this field of research. Just like the others, this volume is intended to provide coverage of some of the highlights of contemporary phosphorus chemistry chosen from the entire range of current interest. Several aspects are emphasized: the chemistry of low coordinated neutral or cationic phosphorus systems, the synthesis of new chiral phosphorus ligands and their use in asymmetric catalysis, the reactivity of primary phosphines or phosphorus ylides, the preparation of imido analogues of phosphorus oxo and chalcogenido anions, and the versatile behavior of difunctionalized phosphorus and sulfur compounds. Profound insight into phosphorus chemistry is indeed given.

Research in the field of catalytic asymmetric reactions is still vigorously growing and phosphorus derivatives occupy a central position in this hot topic. In volume 1, Buono et al reported the use of several chiral organophosphorus catalysts. The first chapter of this volume, written by K.V.L. Crepy and T. Imamoto, is concerned with the synthesis and the applications of other new optically active phosphine ligands mainly obtained from P-boranes derivatives which have emerged as indispensable intermediates. These methodologies provide efficient alternatives to the difficult synthesis of a variety of P-chiral phosphine ligands. Asymmetric transition metal catalyzed reactions with very high enantioselectivity are presented thus allowing P-chiral phosphine ligands to range amongst the leading ligands for the preparation of enantiomerically pure compounds.

The second chapter (M. Taillefer and H.J. Cristau) is dedicated to new trends in ylide chemistry. The preparation and the reactivity of phosphorus ylides, C-substituted by heteroatoms is presented, ylides being substituted by groups 1 and 2 elements, by transition metals or by elements of groups 13 to 16. A rich and versatile chemistry is thus reported.

The third chapter (W.W. Schoeller) reviews the latest developments in the area of donor-acceptor complexes of low coordinated cationic π-bonded phosphorus systems, a topic which attracts the interest of many researchers, both from theoretical and experimental points of view.

The following chapter concerns another kind of low-valent organophosphorus compounds, namely phosphinidenes. Little is known about free phosphinidenes in contrast to the corresponding transition metal complexes . Many new reagents have been generated exhibiting either electrophilic or nucleophilic properties. The reactivity of these carbene-like reagents is evaluated (K. Lammertsma).

Chapter 5 discusses recent developments in the synthesis and properties of primary phosphines. The utility of bromo and aminopropyl phosphines as well as that of carboxylate functionalized primary bisphosphines, the latter for incorporation onto peptides and for their potential applications in catalysis, is underlined by K. V. Katti, N. Pillarsetty and K. Raghuraman.

The chemistry of homoleptic polyimido and heteroleptic imido-oxo- or -chalcogenido anions of phosphorus is described in chapter 6 (T. Chivers). Emphasis is placed on the versatile coordination behavior of these multidentdate ligands, which have both hard (NR) and soft (S, Se, Te) centers.

The last chapter (M. Gulea, S. Masson) is devoted to recent results concerning the synthesis and reactivity of organocompounds substituted by both phosphorus and sulfur functional groups. The association between the phosphonate and dithioester functions induces a very versatile reactivity giving access not only to a variety of phosphorylated heterocycles but also to a number of new (α-sulfanylalkyl) or (β-sulfanylaryl) phosphonates, phosphine oxides or phosphorodiamidates in which chirality can be introduced on phosphorus, on the α carbon, or on sulfur. The use of phosphonothioformates as radical trapping agents and as α RAFT reagents for controlled polymerization certainly opens new perspectives of applications for such dithioesters.

We all hope that you the readers will find this collection interesting to read and a useful source of information for further studies and rewarding applications.

Toulouse, January 2003 Jean-Pierre MAJORAL

Contents

Contents of Volume 220
New Aspects in Phosphorus Chemistry I

Volume Editor: Jean-Pierre Majoral

ISBN 3-540-42246-3

Contents of Volume 223
New Aspects in Phosphorus Chemistry II

Volume Editor: Jean-Pierre Majoral

ISBN 3-540-42246-3

Top Curr Chem (2003) 229: 1–40
DOI 10.1007/b11148

New P-Chirogenic Phosphine Ligands and Their Use in Catalytic Asymmetric Reactions

Karen V.L. Crépy · Tsuneo Imamoto

Department of Chemistry, Faculty of Science, Chiba University, Yayoi-cho, Inage-ku, Chiba 263–8522, Japan. *E-mail: imamoto@faculty.chiba-u.jp*

This chapter is concerned with an update on the preparation and applications of optically active phosphine ligands which bear the chiral center at a phosphorus atom. A great deal of P-chirogenic ligands have been reported over the past ten years, and their unique structure reproduced here. However, only syntheses of ligands which have been the subject of several literature reports in terms of applications to asymmetric catalysis have been summarized. This review also aimed to introduce the latest results in the field of enantioselective reactions.

Keywords. P-Chirogenic phosphine ligands, C_2-symmetry, Phosphine-boranes, Transition-metal catalysts, Catalytic asymmetric reactions

List of Abbreviations

Ad	1-Adamantyl
Ar	Aryl
BINAP	2,2′-Bis(diphenylphosphino)-1,1′-binaphthyl
BisP*	1,2-Bis(alkylmethylphosphino)ethane
Bn	Benzyl
BPE	1,2-Bis(phospholano)ethane
BSA	*N*,*O*-Bis(trimethylsilyl)acetamide
t-Bu	*tert*-Butyl
n-BuLi	Normal butyllithium
s-BuLi	Secondary butyllithium
CAMP	2-Methoxyphenyl(methyl)cyclohexylphosphine
COD	Cyclooctadiene
Cy	Cyclohexyl
DABCO	1,4-Diazabicyclo[2.2.2]octane
DBT	Dibenzoyltartaric acid
de	Diastereomeric excess
DIOP	2,3-*O*-Isopropylidene-2,3-dihydroxy-1,4-bis(diphenylphosphino)-butane
DIPAMP	1,2-Bis[(2-methoxyphenyl)phenylphosphino]ethane
DME	1,2-Dimethoxyethane
DMF	Dimethylformamide
DOPA	3,4-Dihydroxyphenylalanine
DuPHOS	1,2-Bis(phospholano)benzene
E	Electrophile
ee	Enantiomeric excess
Et	Ethyl

HPLC	High performance liquid chromatography
L*	Chiral ligand
LDBB	4,4′-Di-*tert*-butylbiphenyl
m-CPBA	*m*-Chloroperoxybenzoic acid
Me	Methyl
Men	Menthyl
MiniPHOS	Bis(alkylmethylphosphino)methane
MOP	Monodentate phosphine
nbd	Norbornadiene
Nu	Nucleophile
PAMP	2-Methoxyphenyl(methyl)phenylphosphine
Ph	Phenyl
rt	Room temperature
TangPHOS	(1*S*,1*S*′,2*R*,2*R*′)-1,1′-Di-*tert*-butyl-[2,2′]-diphospholanyl
Tf	Trifluoromethanesulfonyl (triflyl)
THF	Tetrahydrofurane
TMEDA	*N*,*N*,*N*′,*N*′-Tetramethyl-1,2-ethylenediamine
TON	Turnover number

1 Introduction

Optically active compounds must be endowed with suitable functionality, configuration, and conformational rigidity or flexibility to produce the desired stereoselectivity, and engender the required biological responses and physical functions. Therefore, discovery of truly efficient methods to obtain chiral molecules is a substantial challenge for synthetic chemists. Asymmetric induction by the use of a chiral catalyst is recognized as being a valuable strategy since only catalytic amounts of the chiral mediator are required, offering obvious economic and practical advantages. Due to the fact that the chiral information in the organic products prepared by enantioselective catalysis derives from the optically active ligands bound to the transition metal, the design of these ligands constitutes a meticulous research. However, relatively few catalysts giving both high enantiomeric excess and accepting a wide range of substrates are available. Therefore, much effort by synthetic organic chemists over the course of the past three decades have been focused on the design of new chiral catalysts or the tuning of existing ligands to meet almost perfect selectivity, excellent reactivity, and high productivity.

A number of excellent reports which deal with synthesis of optically active phosphine ligands are available to date, and have been referenced in this chapter. Therefore it is not the intention here to overlap with them, but rather to describe recent advances in the field. Thus, this chapter is intended to serve as a review to the preparation of some efficient P-chirogenic compounds which have been developed over the past ten years, by either resolution or asymmetric synthesis. Considerable progress has been made in the preparation and use of P-stereogenic compounds. Use of newer methods is stressed here; however an attempt is also made to provide an overview. Section 2 of this chapter comprises a brief survey of chiral phosphines ligands, including phosphine oxides and phosphine-boranes. Section 3 is concerned with preparation of P-chirogenic phosphines. Section 4 describes recent applications of P-chirogenic ligands in transition-metal catalyzed stereoselective reactions.

2
P-Chirogenic Phosphine Ligands

2.1
Generalities

2.1.1
Types of Ligands

A comparative literature study led to the observation that the most effective catalysts present some similar featural requirements. Indeed, it is almost universally observed that ligands with C_2 symmetry elements perform excellent stereochemical control owing to the reduced number of conformations the ligand can assume in the coordination sphere of the metal (ordered environment) – an effect beneficial for the optical induction on product formation [1, 2]. For the same reason, ligands used in enantioselective catalyses are usually bidentate. In some occasions, however, there are advantages in using monodentate ligands [3, 4]. Indeed, there exist transition-metal catalyzed reactions where the bidentate-metal complexes cannot be used because of their low catalytic activity and/or low selectivity toward a desired reaction pathway. Moreover, chelating bidentate ligands are incompatible with the catalytic reactions where the metal catalyst can provide only one coordination site during some steps in the catalytic cycle. Thus, powerful chiral monodentate ligands are sometimes required for the realization of a high level of enantioselectivity.

Finally, although other ligand types continue to be employed in some enantioselective catalysis, optically active phosphines play a key role as chiral ligands in various transition metal-catalyzed asymmetric synthesis, such as hydrogenation, hydrosilylation, hydrocarbonylation, C-C coupling reactions, and isomerization [5, 6], because of their steric and electronic variability. The chelation of the ligand on the metallic center can also occur through a phosphorus atom and another binding group such as sulfide or amine. The two donors atoms play different roles from each other, and the principal advantages in using such chiral hybrid ligands lie in high structural tuning in both steric and electronic means [7–10]. Such structural diversity realizes various reactivity and stereoinduction ability.

In the ligand design, it is important to keep a suitable distance between the two chelating atoms in order to generate a five- to eight-membered ring with the transition metal. If the ring is too large, the ligand tends to bind in a monodentate fashion or form a bridge between the metals [11]. Moreover, it has been reported that sterically congested phosphine ligands remarkably increase the catalytic activities of the metal complexes [12–17].

In the skeleton of many chelating diphosphines, the phosphorus atoms bear two aryl substituents, not least because the traditional route to this class of compounds involves the nucleophilic substitution with alkali metal diarylphosphides of enantiopure ditosylates derived from optically active natural precursors, approach which is inapplicable to the preparation of P-alkylated analogs. The correct orientation of these aryl substituents in the coordination sphere has been identified as a stereochemically important feature contributing to the "recognition ability" of the metal complex [11, 18–20].

The numerous chiral phosphine ligands which are available to date [21] can be subclassified into three major categories depending on the location of the chiral center: ligands presenting axial chirality (e.g., BINAP **1** and MOP **2**), those bearing a chiral carbon-backbone (e.g., DIOP **3**, DuPHOS **4**), and those bearing the chiral center at the phosphorus atom (e.g., DIPAMP **5**, BisP* **6**), as depicted in Fig. 1.

Several fall into the third category, not least because their synthesis presents challenging hurdles. The chiral characteristic encountered in P-chirogenic ligands is however extremely interesting given the direct binding of the chiral atoms to the metal atom. This factor eliminates potentially inefficient secondary trans-

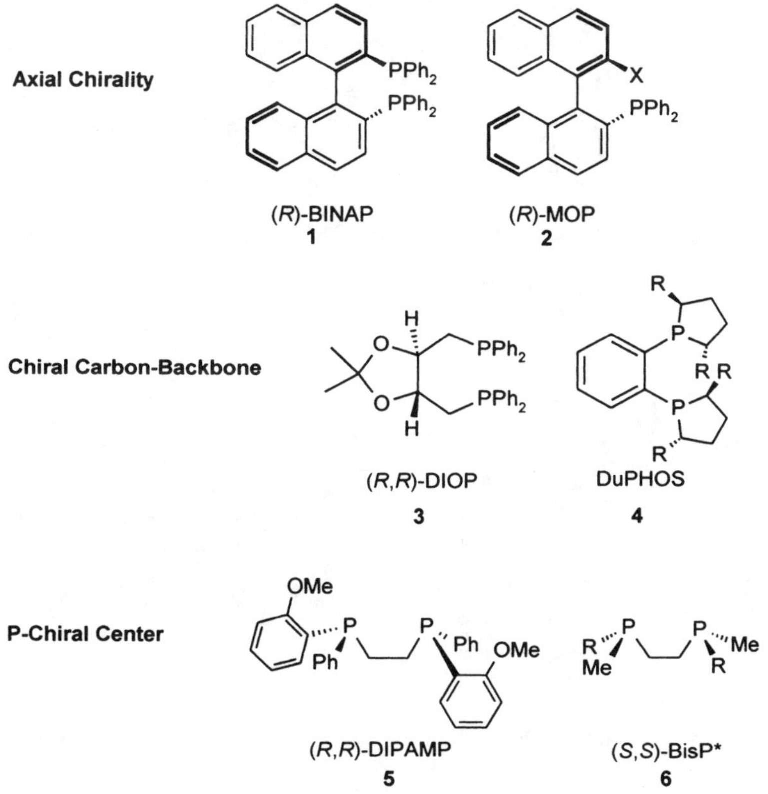

Fig. 1. The three main categories of chiral ligands

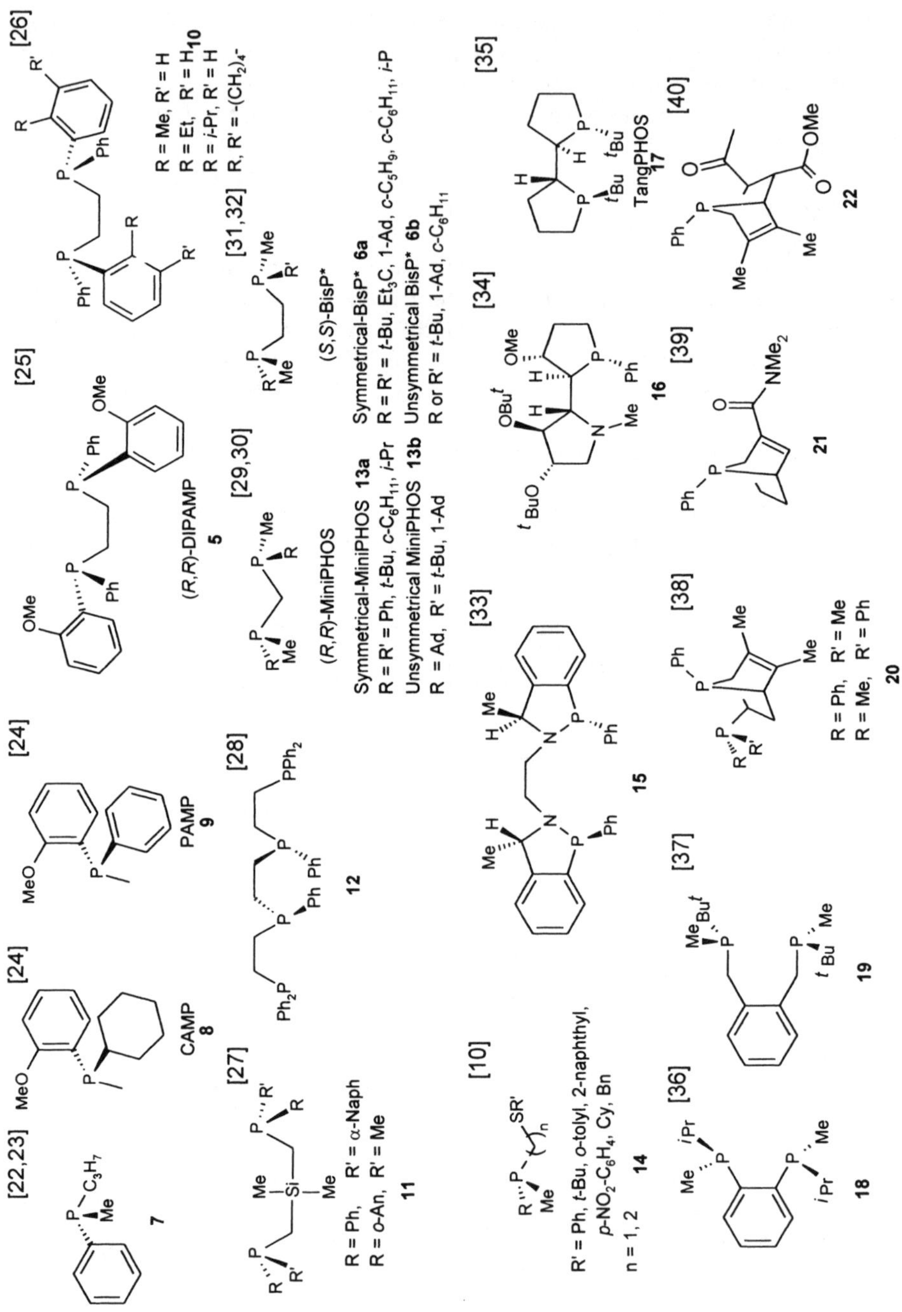

Fig. 2. P-Chirogenic ligands synthesized since 1990 (Ligands **5**, **7**–**9** are indicated here only for reference) [22–53]

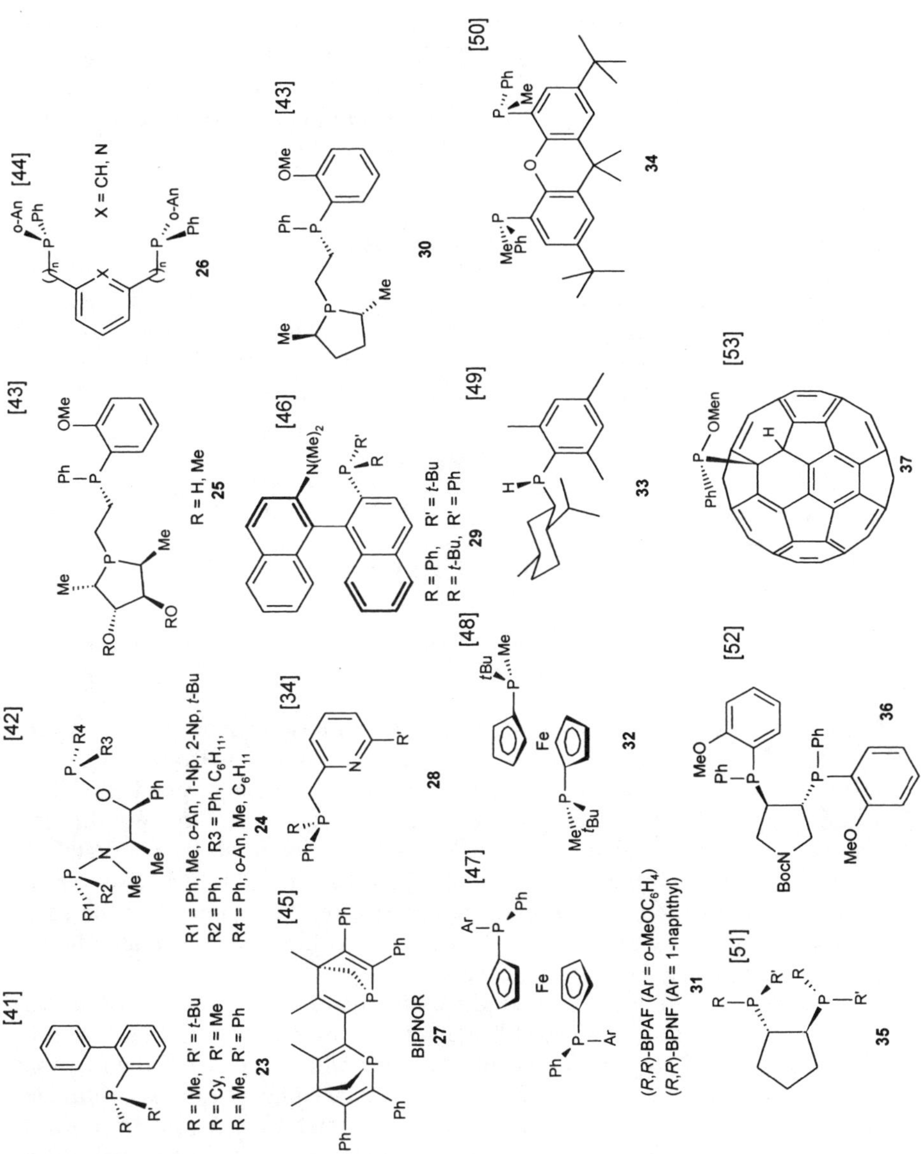

Fig. 2 (continued)

fer of chirality from the ligand backbone of the majority of chiral phosphines and, thus, provides a more effective chiral environment at the site where the enantioselectivity takes place. The most impressive P-chirogenic phosphines synthesized to date are listed in Fig. 2, with their common names and a reference to the paper where the preparation was described or quoted.

2.1.2
Brief Historical Overview of P-Chirogenic Phosphine Ligands

In 1968, Horner et al. [22] and Knowles and Sabacky [23] independently demonstrated that low but definite enantiomeric excesses (up to 15% ee) were produced in the rhodium-catalyzed asymmetric hydrogenation of simple alkenes using methylpropylphenylphosphine **7** as chiral ligand (Scheme 1).

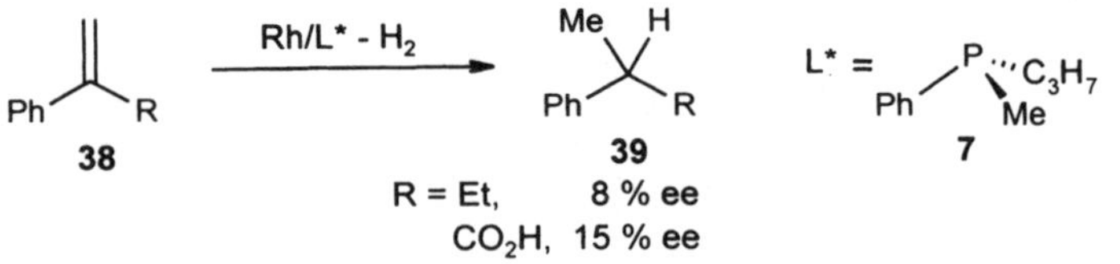

Scheme 1. First use of P-chirogenic phosphine ligands in asymmetric hydrogenation reaction

In the early 1970s, P-chirogenic monophosphines, such as CAMP **8** and PAMP **9** [24], were designed, and ee up to 90% were demonstrated in the reduction of *N*-acetyl dehydrophenyl alanine. In 1975, the C_2-symmetric P-stereogenic diphosphine DIPAMP **5**, discovered by Knowles et al. [25] was the first P-chirogenic ligand to exhibit exceptional activity and high selectivity in catalytic asymmetric hydrogenation of α-(acylamino)acrylic acids [54–56]. Discovery of DIPAMP enabled the first industrial asymmetric synthesis, and thus the commercial process for the amino acid (*S*)-DOPA (3,4-dihydroxyphenylalanine), a drug used for treating Parkinson's disease.

Thereafter, however, P-chirogenic phosphine ligands were the subject of less investigation since the synthesis of highly enantiomerically enriched P-stereogenic phosphines often proves difficult. Another reluctance lies in the fact that this class of phosphines, especially diaryl- and triarylphosphines, is conformationally unstable and gradually racemize at high temperature [57, 58]. In contrast, optically active trialkylphosphines are known to be optically stable even at considerably elevated temperature.

Significant advance in the field of asymmetric catalysis was also achieved with the preparation of 1,2-bis(phospholano)benzene (DuPHOS **4**) and its conformationally flexible derivative (1,2-bis(phospholano)ethane, known as BPE) by Burk et al. [59]. Two main distinctive features embodied by these ligands, as compared to other known chiral diphosphine ligands, are the electron-rich character of the phosphorus atoms on the one hand and the *pseudo*-chirality at phosphorus atoms, on the other. These properties are responsible for both the high activity of the corresponding metal complex and an enantioselection independent of the conformation properties of the chelate cycle [59]. Exceedingly high ees were achieved in the hydrogenation of a wide selection of unsaturated systems [60]. Nevertheless, these ligands present only *pseudo*-chirality at phosphorus atoms and thus will not be included in this chapter.

On the basis of the conclusions accumulated over the years, novel highly efficient P-chirogenic phosphines were designed, synthesized, and tested in a variety of enantioselective reactions. These issues will be stressed in the following paragraphs.

2.2 Optically Active Phosphine Oxides

2.2.1 *Synthesis*

Synthesis of optically active phosphine oxides has been well-documented [58]. Among them, phosphine oxides **40** were one of the first key intermediates in the synthesis of bidentate phosphine ligands (Scheme 2). They are usually synthesized by the reaction of diastereomerically pure menthyl methylphenylphosphinate with Grignard reagents [61] or organolithium reagents [62, 63]. Yet this methodology limits the range of substrates, and the level of optical yield varies with the method employed. We developed a versatile method for the construction of tertiary phosphine oxides via fractional recrystallization of diastereomeric intermediate menthyl 2-phosphinylacetates **44** (synthesized from the corresponding unsymmetrical secondary phosphine oxides **43**) [64–66], in exceedingly high enantiomeric purity (Scheme 2) [67]. The subsequent hydrolysis and decar-

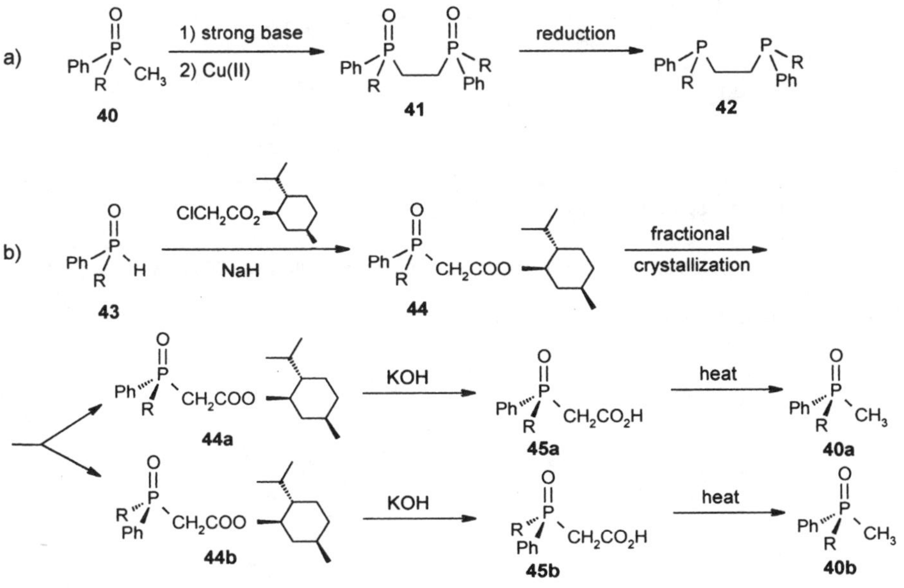

Scheme 2a, b. Efficient route to enantiomerically pure phosphine oxides boxylation proceeded smoothly without any racemization. This method favorably competed with the other reported procedures.

2.2.2
Reduction to Free Phosphines

Stereoselective reduction of optically active phosphine oxides enables isolation of the free phosphines which will be complexed with a transition-metal to yield chiral catalysts for use in enantioselective reactions. Silane reagents were the reducing agents originally used [68–71], but in most cases the transformation is accompanied by partial racemization of the products. $LiAlH_4$ is also a powerful reducing agent, leading to racemized compounds (except in the presence of an activator), owing to the *pseudo*-rotation of the pentacoordinate intermediates [50, 72, 73]. On the other hand, when methyl triflate is used as an activator in DME and left reacted at –60 °C for 2 h prior to addition of $LiAlH_4$, the reduction occurs in high yield with almost net inversion of configuration at the phosphorus atom for a wide range of enantiomerically pure phosphine oxides (Scheme 3) [74]. This methodology constitutes a powerful tool. For example, (*R*)-2-methoxyphenyl(methyl)phenylphosphine, most commonly known as (*R*)-PAMP **9**, was produced in 98% ee (85% yield) by stereoselective reduction of (*R*)-2-methoxyphenyl(methyl)phenylphosphine oxide [74]. Reduction of trialkylphosphine oxides also proceeds smoothly, although chemical yields are less satisfactory.

From mechanistic view points, as represented in Fig. 3, it is likely that the phosphine oxide is methylated by methyl triflate. The resulting phosphonium salt

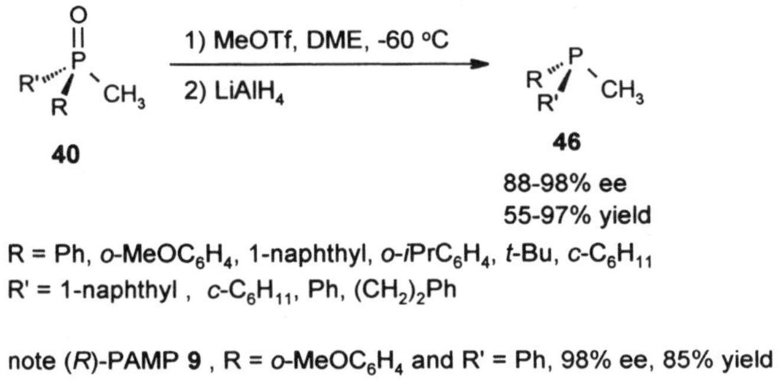

Scheme 3. Stereoselective reduction of phosphine oxides

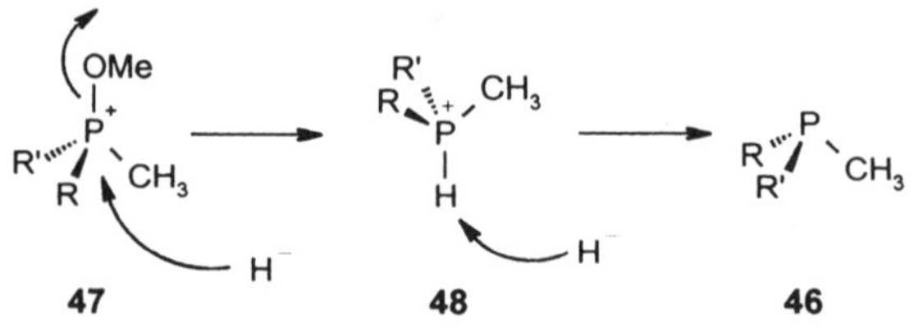

Fig. 3. Proposed mechanism for stereoselective reduction of phosphine oxides

47 is subjected to hydride attack from the backside of the methoxy group to give the reduction product of inverted configuration.

Although the success of this methodology constitutes an improvement over the previously employed protocols, it has become common practice to synthesize phosphine-boranes instead of the corresponding phosphine oxides. This subject is the focus of Sect. 2.3. Note, however, that attempts to convert enantiomerically enriched tertiary phosphine oxides directly into the corresponding phosphine-boranes have been shown to proceed in good yields by reaction with $LiAlH_4$, $NaBH_4$, and $CeCl_3$ [73]. Unfortunately, the phosphine-borane produced often is a racemate, due to the use of $LiAlH_4$. Cerium chloride is believed to coordinate to the phosphine oxides, facilitating the deoxygenation with $LiAlH_4$, and to coordinate $NaBH_4$ to transform the intermediate phosphines into phosphine-boranes.

2.3
Phosphine-Borane Complexes as Stable Intermediates

Tricoordinate phosphorus compounds in low oxidation states are usually air-sensitive, making their handling and storage difficult. Moreover chiral phosphines bearing stereogenic phosphorus are prone to racemization especially at high temperature [57, 58], and therefore require additional stabilization. Temporary protection of the phosphorus with BH_3 is in most cases the solution to these problems [75, 76]. The phosphine-boranes obtained are air-stable compounds, and hence they can be conveniently isolated, purified, and stored. Furthermore, owing to their remarkable inertness and resistance to a wide range of reactions conditions P-boranes have emerged as indispensable intermediates for the preparation of P-chirogenic phosphines [77].

Besides its protective function of the labile phosphine group, the BH_3 group activates the adjacent substituents such as methyl group or P-H bond to deprotonation with a strong base [78]. This methodology provides an efficient alternative to the difficult synthesis of a variety of optically active tertiary phosphine derivatives, as will be described in Sect. 3.

2.3.1
Boranation – Deboranation

Boranation of P-stereogenic phosphines is readily and quantitatively performed from the corresponding free phosphines by treatment with, usually, BH_3.THF complex, and occurs with complete preservation of the stereochemical integrity at phosphorus (Scheme 4). Liberation of the enantiomerically pure phosphines is achieved from the corresponding borane complexes in a stereospecific manner with retention of configuration, using two main methods (Scheme 4). The first consists of the cleavage of the P-B bond by treatment with an excess of amine which has strong nucleophilicity [73, 78]; pyrrolidine, morpholine, or DABCO are frequently employed [73, 79]. The second method is due to the research undertaken by Livinghouse et al.. They demonstrated that removal of more inert P-B bonds of electron-rich phosphines is most effectively accomplished by the reaction with strong acids such as CF_3SO_3H or HBF_4–OMe_2, followed by treatment

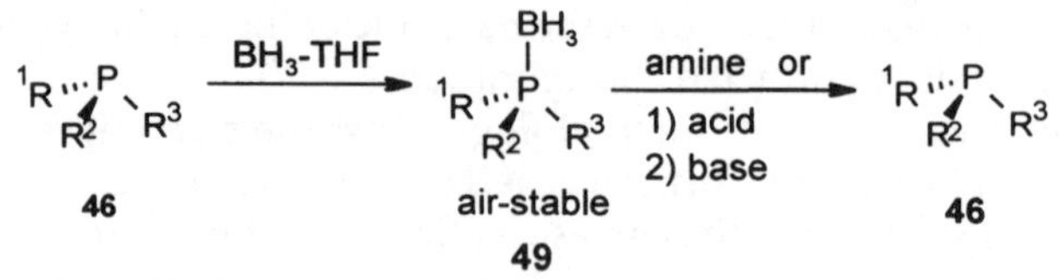

Scheme 4. Boronation/deboranation of phosphines

with NaOH or $NaHCO_3$ [80, 81]. It should be noted that both methods are carried out under mild conditions and hence, in most cases, the stereochemical integrity or stereogenic phosphorus remains intact [73].

2.3.2
Purification – Complexation

After deboranation it is usually necessary to purify the free phosphine before treating it with a transition-metal complex to produce a transition-metal/phosphine ligand catalyst. In the case of deboranation using an amine, simple evaporation of the unreacted amine (which also serves as the solvent) precedes the purification. On the other hand, extraction with a degassed organic solvent in argon atmosphere is required when the phosphine complex has been treated with an aqueous base. The organic extract obtained is subsequently dried over common drying agents, and the filtered solution evaporated. In both cases, the residue obtained from evaporation is passed through a short column of basic alumina or silica gel eluting with degassed solvent under an inert atmosphere. After removal of the solvent, the purified free phosphine is immediately complexed with a stable (commercially available or readily synthesized) transition-metal complex in a degassed solvent. These two species combine to give the active catalyst within minutes. The catalyst formed is either isolated or used in situ. After addition of other substrates, the asymmetric reaction takes place. One advantage of in situ preparation is the possibility of changing the metal-to-ligand ratio. On the other hand, isolated catalysts sometimes provide better chemical or optical yields. They also enable unambiguous mechanistic studies, where questions may remain unsolved in the case of catalyst generated in situ.

3
Routes to P-Chirogenic Phosphines

3.1
The Great Potential of the P-OMen Functionality

3.1.1
Synthesis and Structural Modifications

It is now well-established that unsymmetrical substituted menthylphosphinates, RR′P(O)OMen, as well as the corresponding phosphine-boranes RR′P(BH_3)OMen and phosphinothioates RR′P(S)OMen, can be separated readily into the diastereomeric forms, and subsequently reacted with Grignard reagents to afford P-chirogenic tertiary phosphines with a high degree of stereospecificity [57]. The pioneering work in this area is due to Nudelman and Cram [82], and to Mislow and co-workers [61, 83]. The majority of these menthyl intermediates are prepared by successive treatments of dichlorophenylphosphine with (–)-menthol and other electrophiles. Displacement of the OMen group by organometallic reagents occurs with inversion of configuration at phosphorus. For example, the reaction of methyllithium and diastereomerically pure *l*-menthyloxy(*o*-anisyl)phenylphosphine-borane **50** produced optically active *o*-anisylmethylphenylphosphine-borane **51** in high yield with inversion of configuration and without any loss of enantioselectivity (Scheme 5a) [73]. It was demonstrated that the *o*-methoxy group remarkably enhances the rate in the nucleophilic substitution reaction thanks to the coordination of the methoxy oxygen to the lithium ion. In other words methyllithium is brought more closely to the reaction site [84].

We discovered a complementary procedure for conversion of OMen to other functional groups. The ester P-OMen bond was shown to be cleaved in a stereoselective manner reductively [85, 86]. The cleavage takes place with almost complete preservation of stereochemical integrity at phosphorus. The reducing agents are usually sodium or lithium naphthalenide, lithium biphenylide, and lithium 4,4′-di-*tert*-butylbiphenyl (LDBB). The species produced is then quenched with an alkyl halide or methanol to afford tertiary or secondary phosphines, respectively (Scheme 5b). Overall, the displacement reaction proceeds with retention of configuration.

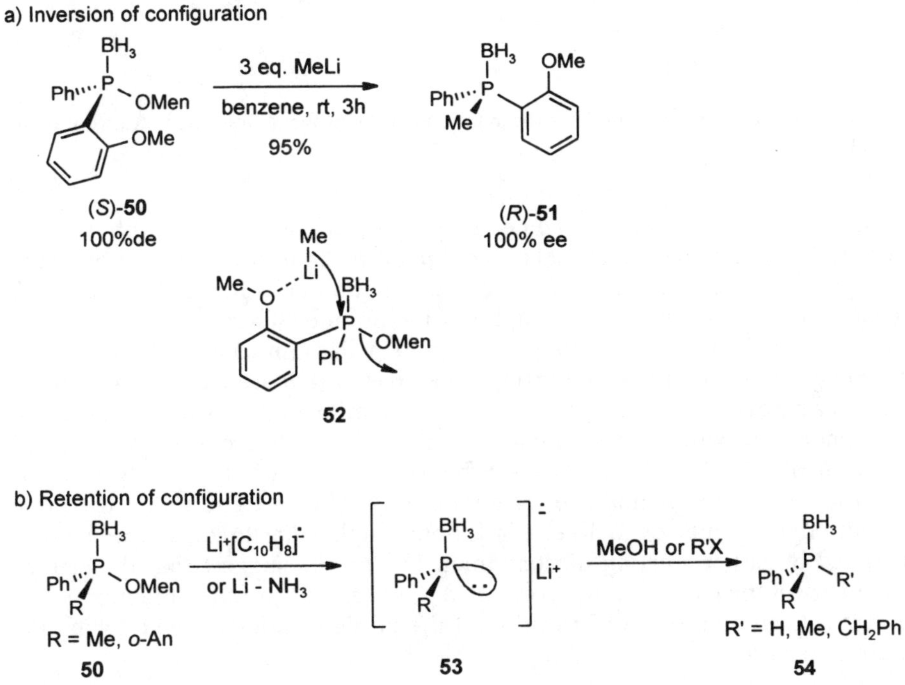

Scheme 5 a, b. Syntheses of P-chirogenic intermediates from menthylphosphines

For example, precursor (*S*)-**57** may be prepared by reaction of (*R*)-**56** with methyllithium or by reduction of (*S*)-**56** with lithium naphthalenide and subsequent methylation with iodomethane, making the overall transformation of the diastereomeric mixture **56** nearly quantitatively (Scheme 6) [87].

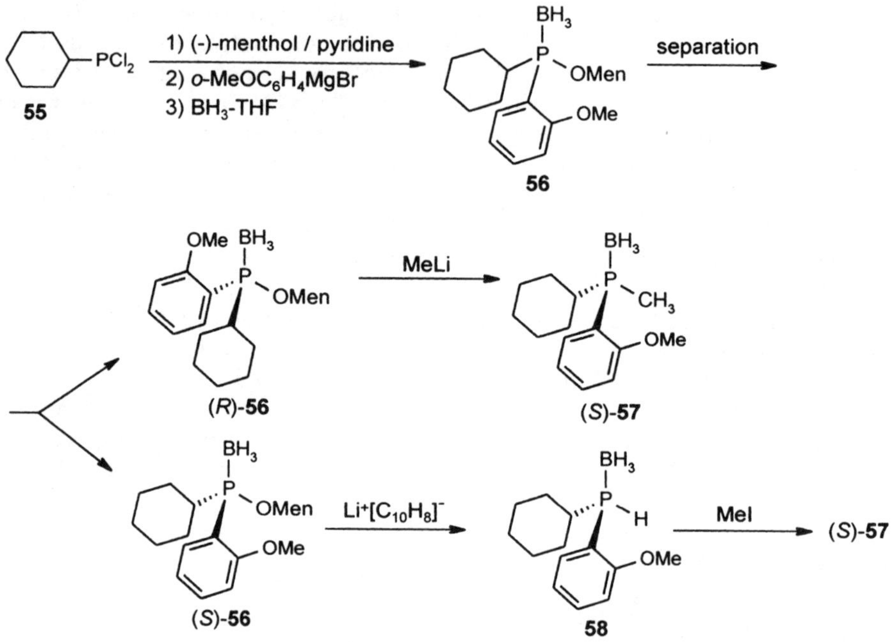

Scheme 6. Representative example of the synthesis of P-chirogenic intermediates from menthylphosphines

Synthesis employing diastereomerically pure menthyloxy(phenyl)phosphine-borane **60** as the starting material is worth pointing. Indeed, deprotonation of the (S_P)-precursor **60** and quenching of the anion obtained with methyl iodide afforded (menthyloxy)methylphenylphosphine-borane **50** with retention of configuration (Scheme 7a) [73]. Interestingly, the palladium-catalyzed ($Pd[PPh_3]_4$) coupling reaction of the same starting material **60** with *o*-iodoanisole in the presence of a base occurred either with complete retention of the configuration of the phosphorus or with almost complete inversion depending on the solvent used (acetonitrile and THF, respectively. Scheme 7b) [88]. Reaction using the (*R*)-enantiomer of the starting substrate followed the same trend under the same conditions (Scheme 7c). In other words both coupled diastereomers can be synthesized by either starting diastereomer. The results suggest that the stereochemistry of the reaction is not controlled by the chirality of the *l*-menthyl group and it is likely that the configuration of the product is determined during the transmetallation step in the catalytic cycle [88].

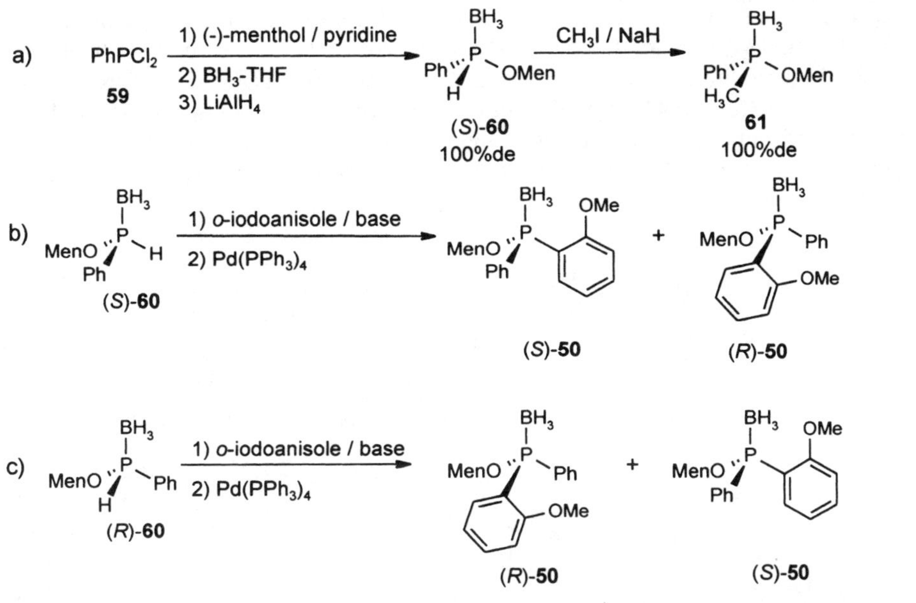

solvent	time(h)	yield	(*R*)-**50** : (*S*)-**50**
CH_3CN	16	98	100 : 0
THF	48	67	4 : 96

Scheme 7 a–c. Switch in enantioselectivity with the reaction conditions

3.1.2
Examples

Ligands **23** which are represented in Scheme 8b–d serve as a synthetic example of the above facts [41]. Monodentate phosphines possessing 2-biphenyl group have been employed as effective ligands in asymmetric catalyses such as Pd-catalyzed carbon-carbon bond forming reactions [13–16]. Their key precursors **62** were synthesized from the reaction between the corresponding alkyl- or phenyldichlorophosphine, 2-lithiobiphenyl, lithium *l*-menthoxide, and borane-THF to afford diastereomer mixtures (ca. 1:1) in 62–74% yields (Scheme 8) [41]. Pure diastereomers were obtained by separation by preparative HPLC or recrystallization, and their structure unequivocally assigned by single crystal X-ray analyses.

Phosphine-borane **63a** (75% ee) was obtained by reduction of compound (S_P)-**62a** using LDBB at –60 °C and nucleophilic substitution with iodomethane in 72% yield. The observed loss of optical purity may be ascribed to stereomutation of the generated tricoordinated phosphorus species. Recrystallization afforded (*S*)-**63a** in >99% ee. On the other hand, severe racemization was observed using the same method with (R_P)-**62b**. An alternative strategy consisted of deboranation of (R_P)-**62b** using *N*-methylpyrrolidine, methylation with methyl triflate, stereospecific reduction with $LiAlH_4$ [74], and temporary protection employing borane-THF complex to afford (*R*)-**63b** in 97% ee (>99% after recrystallization) and high yield. Finally, (*S*)-**63c** was readily prepared from (R_P)-**62c** by treatment with methyllithium [41]. Usual deboranation reactions employing *N*-methylpyrrolidine were applied to all three ligands to afford (*S*)-**23a**, (*R*)-**23b**, and (*S*)-**23c** which must be immediately complexed with a suitable transition-metal complex for application in asymmetric catalysis.

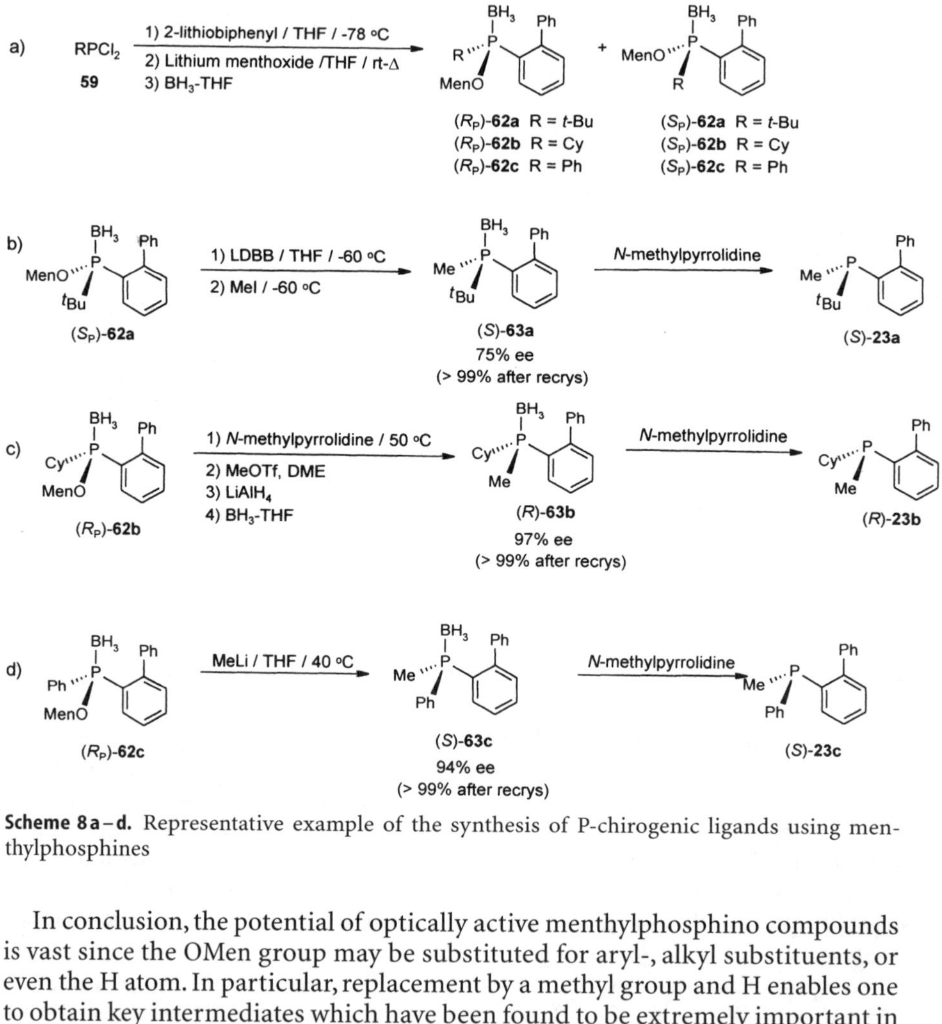

Scheme 8a–d. Representative example of the synthesis of P-chirogenic ligands using menthylphosphines

In conclusion, the potential of optically active menthylphosphino compounds is vast since the OMen group may be substituted for aryl-, alkyl substituents, or even the H atom. In particular, replacement by a methyl group and H enables one to obtain key intermediates which have been found to be extremely important in the synthesis of other substituted monophosphines or diphosphines, as will be shown in the sections which follow.

3.2
P-CH_3 as Key Functional Group

3.2.1
Chiral Starting Phosphines

The methyl group borne on chiral phosphine-borane precursors may be easily deprotonated on treatment with a strong base such as *s*-BuLi [89], permitting the synthesis of a variety of functionalized chiral phosphine-boranes. Indeed, the generated carbanion may be subjected to *C*-alkylation [89] using various electrophiles such as allyl bromide, chlorotrimethylsilane, or carbonyl compounds, and usually high yields of the corresponding monophosphine-borane are produced [73]. Alternatively the same metallated species can undergo a copper(II)-promoted oxidative coupling [90] (without impairment of the borane functionality) to give symmetrical 1,2-bis(boranatodialkyl/arylphosphino)ethane in high yields [73]. Some examples of this method are demonstrated below (Scheme 9).

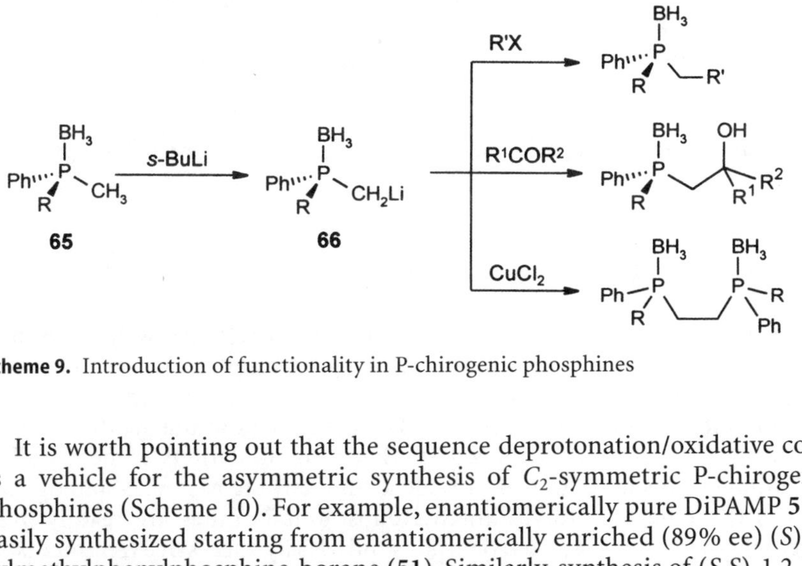

Scheme 9. Introduction of functionality in P-chirogenic phosphines

It is worth pointing out that the sequence deprotonation/oxidative coupling is a vehicle for the asymmetric synthesis of C_2-symmetric P-chirogenic diphosphines (Scheme 10). For example, enantiomerically pure DiPAMP **5** can be easily synthesized starting from enantiomerically enriched (89% ee) (*S*)-*o*-anisylmethylphenylphosphine-borane (**51**). Similarly, synthesis of (*S*,*S*)-1,2-bis[cyclohexyl(*o*-methoxyphenyl)phosphino]ethane (**71**), which structurally resembles DiPAMP, was devised using precursor **70** [87]. Enantiomerically pure (*S*,*S*)-1,2-bis[(*o*-ethylphenyl)phenylphosphino]ethane **10**, which skeleton also is similar to (*S*,*S*)-DiPAMP but possesses no oxygen functional group, was prepared via a similar strategy [26]. The results obtained in the asymmetric hydrogenations of α-(acylamino)acrylic acids catalyzed by the Rh-complex of the latter ligand were comparable to those obtained with DiPAMP (opposite configuration was however obtained), ruling out the speculated argument that the weak binding of the methoxy group borne on DiPAMP with the Rh-metal plays a role in the stereoregulation of the reaction. Steric effects seem to be more important that the coordinative interaction.

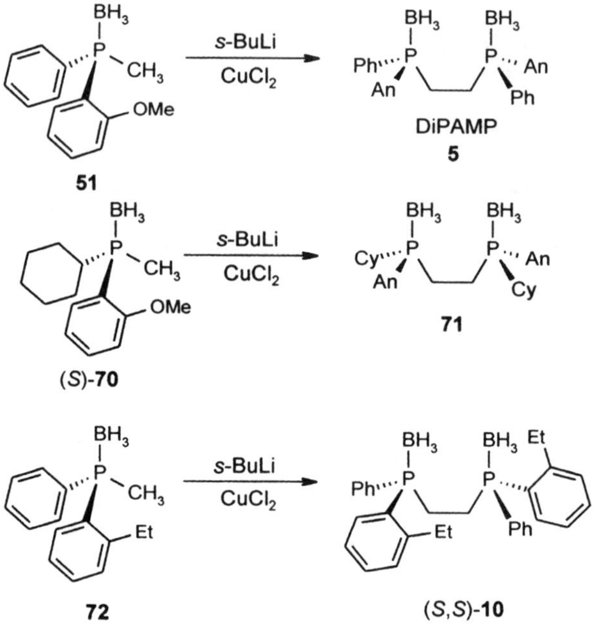

Scheme 10. Representative examples of the synthesis of C_2-symmetric phosphines

3.2.2
Achiral Starting Phosphines

Ever since Evans et al. discovered that prochiral alkyl(dimethyl)phosphine-boranes may undergo enantiodifferentiating deprotonation of one methyl group using butyllithium in the presence of (–)-sparteine as the chiral inductor [27], these compounds have become attractive key precursors to the synthesis of P-chirogenic phosphine-boranes. Indeed, synthetic manipulations of the chirally-induced lithiated species obtained provide a simple route to new P-stereogenic phosphines (Scheme 11). Alkyl(dimethyl)phosphine-boranes are readily prepared in high yields via three-step one-pot synthesis by successive treatments of PCl_3 with alkyl Grignard reagent, methylmagnesium bromide, and BH_3-THF complex. The crude products are recrystallized in suitable solvent to give the de-

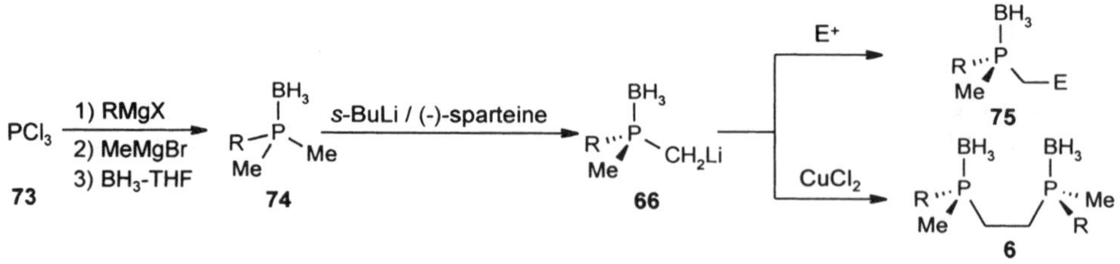

Scheme 11. Enantioselective deprotonation of prochiral phosphines

sired precursors in good yields averaging 75% depending on the nature of the alkyl group. The aim of the following paragraphs is to explore the synthesis of newly developed chiral diphosphine ligands employing this strategy.

3.2.2.1
Example: Introduction of a Heteroatom

Alkyldimethylphosphine-boranes **74** underwent enantioselective deprotonation employing (–)-sparteine/*s*-BuLi, followed by oxidation with molecular oxygen [91, 92] in the presence of triethyl phosphite (Scheme 12) to afford moderate yields of enantiomerically enriched alkyl(hydroxymethyl)methylphosphine-boranes **76**, with 91–93% ee in the case of a bulky alkyl group and 75–81% ee in the case of cyclohexyl or phenyl groups [93]. Except for the adamantyl derivative (in which the ee increased to 99%), no major improvement in the ee was observed after recrystallization.

Quenching of the same lithiated species with CO_2, followed by reduction of the carboxylic acid functionality obtained with BH_3-THF complex, yielded the next higher analogues **78** to these alcohols [94]. Subsequent treatment of the deprotonated alcohols with TsCl or MsCl afforded (*R*)-1-boranato[alkyl(methyl)phosphino]ethanol-2-tosylates or the mesylate phosphine-boranes in over 90% ee and excellent overall yields.

P-Chirogenic phosphine/sulfide hybrid phosphine-boranes **80** were synthesized from the reaction between (*R*)-tosylates **79** [94] and sodium thiolate in DMF at ambient temperature as depicted in Scheme 12, or alternatively by a one pot synthesis consisting of the nucleophilic attack of the chirally induced lithium salt of **74** on phenyl disulfide. Both methodologies provided the desired sulfide/phosphine boranes in excellent yields [10].

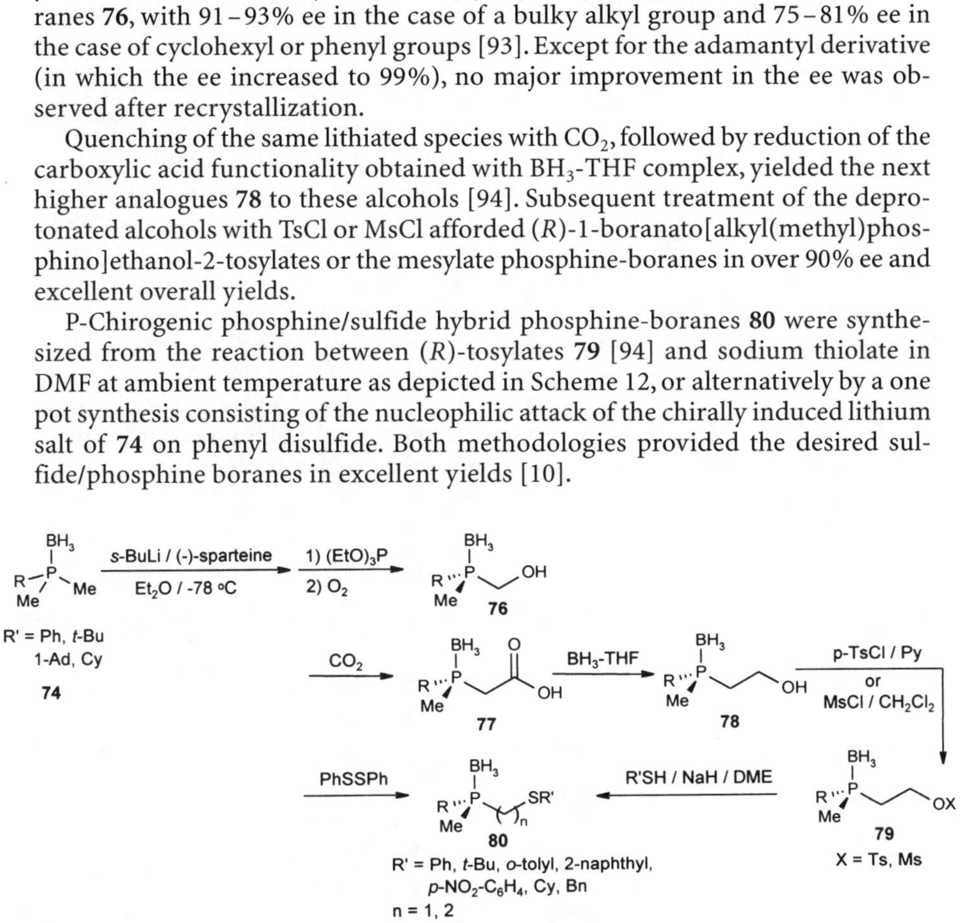

Scheme 12. Introduction of functionality in prochiral phosphines

3.2.2.2
Synthesis of BisP and MiniPHOS*

BisP* (which is the common name for 1,2-bis(alkylmethylphosphino)ethanes **6a**) and MiniPHOS (bis(alkylmethylphosphino)methanes **13a**) are electron-rich, C_2-symmetric P-stereogenic ligands which have been designed and prepared almost four years ago (Scheme 13) [29, 31]. One of the important features of these ligands is that a bulky group and the smallest alkyl group (Me) are borne on each phosphorus atoms. Also, the steric bulk imposed by the substituents borne on the phosphorus atoms may be readily changed without affecting the electronic nature of the metal. Facile variation of the substituents enables optimization of the enantioselectivity by conferring correct steric matching between the ligand and the incoming substrate.

Alkyldimethylphosphine-boranes **74** were subjected to enantioselective deprotonation using *s*-BuLi in the presence of (–)-sparteine [27], followed by copper-promoted oxidative coupling [90] of the generated anion (Scheme 13). Since oxidative coupling proceeds through a radical intermediate, the reaction leading to homochiral isomers ((*S,S*) or (*R,R*)) has almost the same rate as that to produce *meso* isomer (*RS*) [73]. Thus, the diphosphines obtained ((*S,S*)-BisP*-boranes **81a**) were accompanied by a small amount of *meso*-diastereomer (>95% ee for R=*t*-Bu, 1-Ad, Et_3C, and 70–80% ee for R=*c*-C_5H_9, *c*-C_6H_{11}, *i*-Pr) which was readily removed by direct recrystallization or by flash chromatography followed by recrystallization to afford the enantiomerically pure diphosphines [31]. The lower enantiomeric purity observed with some R substituents may be explained by a less selective enantioselective deprotonation of the corresponding starting alkyl(dimethyl)phosphine-boranes. The overall yields ranged from 20 to 40% depending on the alkyl groups borne on the diphosphines.

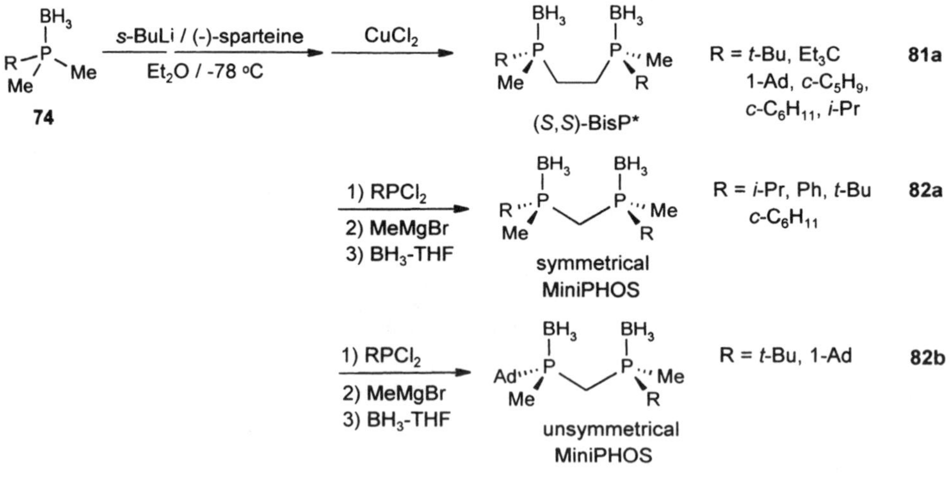

Scheme 13. Representative examples of the synthesis of P-chirogenic ligands from prochiral phosphines

The same phosphine-borane used for the synthesis of BisP* acted as the starting materials of the construction of MiniPHOS, the next smaller analogue to BisP* (Scheme 13). The chirally induced lithium salt was treated with alkylphosphorus dichloride, methylmagnesium bromide, and borane-THF complex to afford enantiomerically pure MiniPHOS-borane **82a**. Recrystallization enabled elimination of a small amount of corresponding *meso*-diastereomer formed [29]. Yields were generally low, ranging from 13 to 28%.

We further synthesized unsymmetrical MiniPHOS derivatives **13b** (Scheme 13) [30]. Thus, enantioselective deprotonation of 1-adamantyl(dimethyl)phosphine-borane (**74**, R=1-Ad), followed by treatment with *tert*-butyldichlorophosphine or 1-adamantyldichlorophosphine, methylmagnesium bromide and borane-THF complex afforded the optically active diphosphine-boranes **82b** as a mixture with the corresponding *meso*-diastereomer. Enantiomerically pure unsymmetrical MiniPHOS-boranes **82b** were obtained by column chromatography on silica gel or separation by recycling preparative HPLC.

3.3
Secondary Phosphine Boranes

3.3.1
P-H as Key Functional Group

Similarly to the P-CH_3 group, secondary phosphine-boranes react smoothly in the presence of a base (BuLi, NaH) under mild conditions to afford other kinds of functionalized phosphine-boranes in good to high yields, without racemization. Yet the success of deprotonation/treatment with an electrophile process to afford substituted phosphine derivatives without any loss in optical purity may depend on the deprotonation agents employed. Use of butyllithium usually provides the products with high enantiomeric excess in good to high yields [73].

The deprotonated species can be involved in a broad range of nucleophilic reactions, using various electrophiles amongst which reactive alkyl halides, epoxides or benzyne are the most frequently used reagents (Scheme 14) [73, 87]. The Michael-type addition of deprotonated secondary phosphines to olefins bearing an electron-withdrawing group also proceeds smoothly in almost perfect ee [73, 87]. Reactions with acid chlorides give mixed acid anhydride possessing a chiral, optically active component, which may serve as potentially useful enantioselective acylation agents [87]. Nucleophile attack on carbonyl groups is possible, although this method is generally limited to aldehydes, strained rings and less sterically crowded ketones [73].

As mentioned in Sect. 3.1.1, secondary phosphine-boranes also react efficiently with aryl iodides in palladium-catalyzed substitution reactions ($Pd(PPh_3)_4$) [73]. In all cases the boranato functional group remains unchanged.

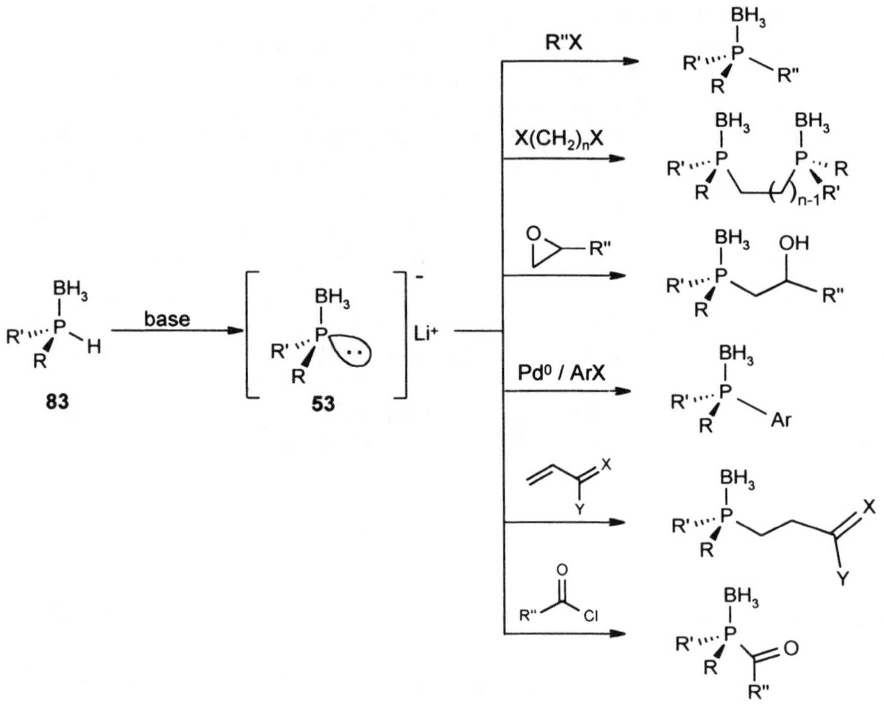

Scheme 14. Introduction of functionality in secondary P-chirogenic phosphines

3.3.2
Examples

Bischirogenic unsymmetric (*S*,*S*)-BisP* **6b** (cf. (*S*,*S*)-BisP*-borane **81b**) were synthesized from the coupling reaction between synthon **79** and the lithiated (*S*)-alkylmethylphosphine-boranes **87** in reasonable to quantitative yields with enantioselectivity over 97% (Scheme 15) [94]. These phosphines constitute the unsymmetric version of BisP* in that they bear different groups on both phosphorus atoms, breaking the C_2-symmetry character of BisP* [32, 94].

P-Chirogenic diphosphine **19**, which rhodium-chelate complex forms a seven-membered ring (rare case for P-stereogenic ligand), was also prepared in reasonable yield (68%) using the wide chemistry of secondary phosphine borane [37]. Deprotonation of the enantiomerically enriched *tert*-butylmethylphosphine-borane **88** (Scheme 15) followed by quenching with α,α'-dichloro-*o*-xylene and recrystallization afforded optically active diphosphine-borane **89** (precursor of free phosphine **19**).

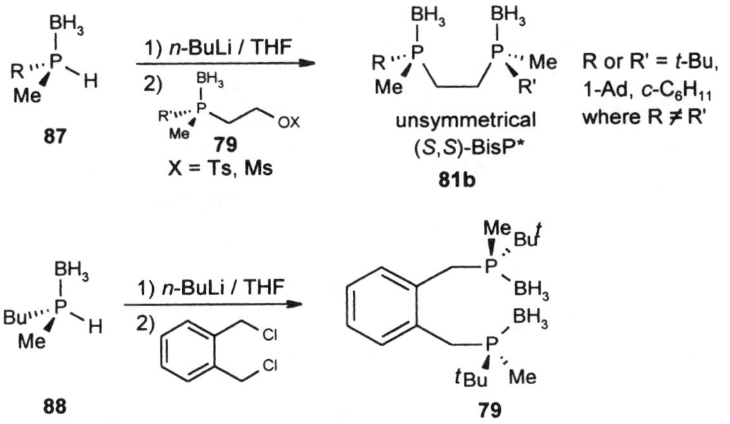

Scheme 15. Representative examples of the synthesis of P-chirogenic ligands from secondary phosphines

3.3.3
Improvement of the Synthesis of P-H Compounds

The majority of newly designed P-chirogenic diphosphine ligands are prepared by the use of naturally occurring (–)-sparteine as a chiral source. However, the synthesis of their counter-enantiomers remains difficult mainly owing to the unavailability of (+)-sparteine as a chiral source. Investigation of the synthesis of optically active secondary dialkyl-substituted monophosphine-boranes **87** in both enantiomeric forms, which serve as key intermediates to both (*R*)- and (*S*)-trisubstituted monophosphines and (*R*,*R*)- and (*S*,*S*)-trisubstituted diphosphines, identified this problem (Scheme 16a) [95]. On the basis that the menthyloxy group of diastereomerically pure (menthyloxy)alkylphenylphosphine borane can be reductively removed to produce optically active alkylphenylphosphine-borane (cf. Sect. 3.1) [85], diastereomeric mixtures of bornylthio-(alkyl)methylphosphine-boranes were synthesized from the corresponding alkyldichlorophosphine by successive electrophilic additions and protection with borane-THF complex (93% combined yield) [95]. Diastereomerically pure **90a** and **90b** were obtained by separation by preparative HPLC. Reduction with lithium naphthalenide and cleavage of the P-S bond at low temperature occurred with virtually net retention of the configuration at the phosphorus atom. Storage of these compounds at ambient temperature for one week constitutes no problem in terms of stereochemical stability. In contrast to the lithiated secondary phosphine-boranes bearing an aromatic substituent, the racemization half-life of the lithiated-**87** species is quite long and calculated to be 7.4 days at 20.0 °C [95], data which is consistent with the results on the racemization of tertiary phosphines [96].

The use of these dialkyl lithiated species enabled the first synthesis of the (*R*,*R*)-counter-enantiomer of (*S*,*S*)-BisP* in almost perfect ee and reasonable yield (Scheme 16b) [31, 95].

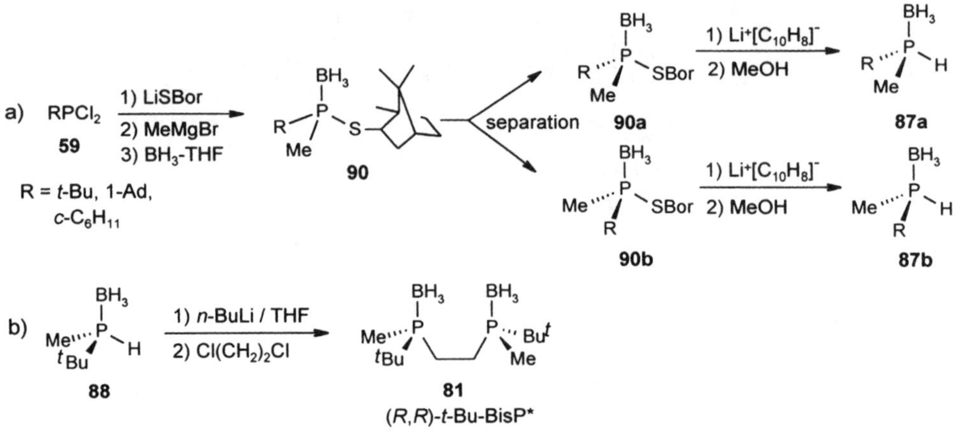

Scheme 16. **a** Production of both enantiomeric forms of secondary phosphines. **b** First synthesis of (*R*,*R*)-BisP*

The advantage that lies in this method is that both enantiomers can simultaneously be prepared. Yet separation of the mixture of diastereoisomers means that 50% is the highest expected yield. Moreover, use of preparative HPLC is not applicable on a large scale. Therefore, an alternative route to the construction of the (*S*)-enantiomer of the same secondary alkylmethylphosphine borane was conceived employing alkyl(hydroxymethyl)phosphine-boranes **76** as the starting materials (Scheme 17a) [93]. The transformation is a one-carbon oxidative degradation of the primary alcohol moiety performed in the presence of hydrated ruthenium trichloride, potassium hydroxide, and potassium persulfate to produce the desired secondary phosphine-borane derivatives **85** in high yields (80–90%), with no degradation of the borane moiety and no racemization. The complex $[Ru(VI)O_4]^{2-}$ generated in situ from $RuCl_3 \cdot x\,H_2O$ in excess aqueous base

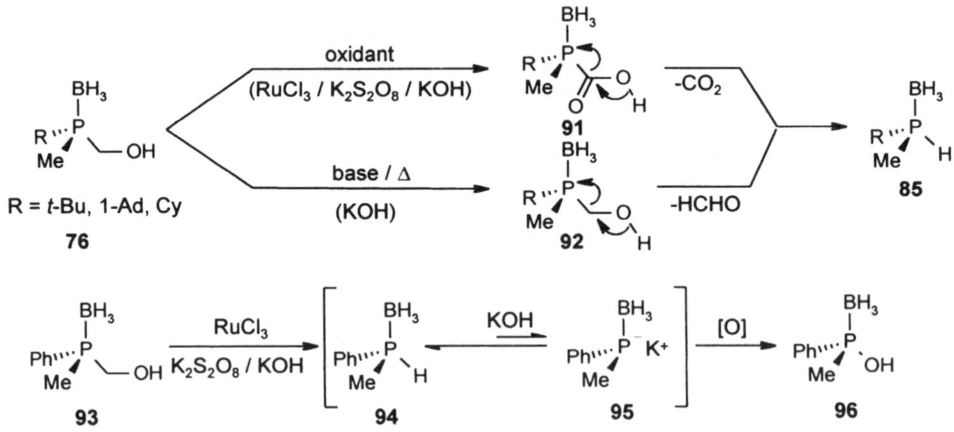

Scheme 17. Improved synthesis of optically active secondary phosphine-boranes

by persulfate is considered to be the oxidation agent [97]. Although this method proves efficient to alkyl(hydroxymethyl)methylphosphine-boranes, overoxidation is observed in the case of the corresponding phenyl(hydroxymethyl)methyl-derivative **93**, and compound **96** is isolated instead of the secondary phenyl-(methyl)phosphine-borane **94**. This can be ascribed to the rather acidic character of the initially formed **94**, suggesting that deprotonation occurs and the generated phosphorus anion oxidizes [93].

Base-promoted deformylation (KOH/Δ) to afford the same secondary phosphine-boranes is less effective and an acceptable compromise between yield and selectivity is difficult to reach. When the reaction proceeds smoothly, it is accompanied by significant racemization [93].

3.4 Examples of Other Interesting Ligands

3.4.1 *P-Chirogenic Diphosphine Oxides*

As mentioned in Sect. 2.2, phosphine oxides are air-stable compounds, making their use in the field of asymmetric catalysis convenient. Moreover, they present electronic properties very different from the corresponding free phosphines and thus may be employed in different types of enantioselective reactions. *m*-Chloroperbenzoic acid (*m*-CPBA) has been showed to be a powerful reagent for the stereospecific oxidation of enantiomerically pure P-chirogenic phosphine-boranes [98], affording (*R*,*R*)-**97** from Ad-BisP* **6** (Scheme 18) [99]. The synthesis of (*R*,*R*)-**98** and (*S*,*S*)-**99**, which possess a *t*-Bu substituent, differs from the precedent in that deboranation precedes oxidation with hydrogen peroxide to yield the corresponding enantiomerically pure diphosphine oxides (Scheme 18) [99].

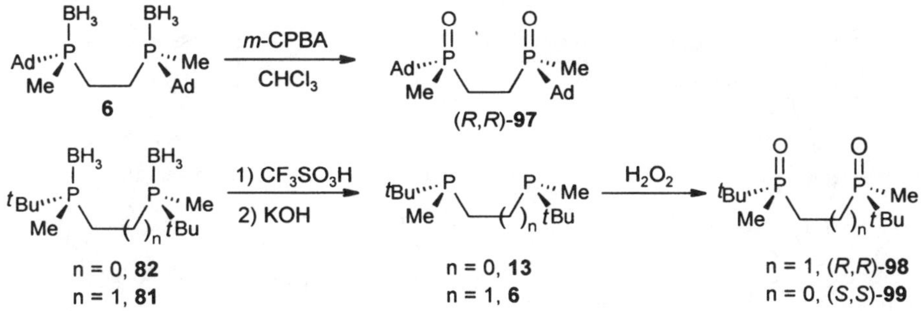

Scheme 18. Example of synthesis of enantiomerically pure phosphine oxides

3.4.2
*(R,R)-BisP**

The combined use of the oxidation of a free phosphine [99] and stereospecific reduction of the phosphine oxide [74] was employed for the synthesis of (*R*,*R*)-*t*-Bu-BisP* from (*S*,*S*)-BisP* itself. Specifically, (*S*,*S*)-BisP* was deboronated (TfOH/KOH), oxidized (H_2O_2), and reduced (MeOTf/$LiAlH_4$) [74], producing, unfortunately, low yields of the desired counter-enantiomer.

3.4.3
Enantiomerically Pure 1,2-Bis(isopropylmethylphosphino)benzene

Burk and co-workers demonstrated that the added rigidity of DuPHOS, as compared to BPE, was translated to excellent enantioselectivity in the rhodium-catalyzed asymmetric hydrogenation of a wide range of N-protected enamide esters and acids with selectivity approaching 100% [59]. 1,2-Bis(isopropylmethylphosphino)benzene **18**, which is a rigid analogue (1,2-phenylene backbone) to BisP*, was therefore tested by us (Scheme 19) [36]. Enantiomerically pure ligand (*S*,*S*)-**18** was prepared starting from 1,2-bis(methylphosphino)benzene **100** [100] by deprotonation using BuLi followed by treatment with 2-bromopropane and subsequent oxidation with hydrogen peroxide. The racemic phosphine oxides obtained were separated from their *meso*-analogues by recrystallization. Optical resolution of (rac)-**101** was achieved using dibenzoyl-*d*- or *l*-tartaric acid ((+)- or (–)-DBT) followed by decomposition with NaOH to afford enantiomerically pure (*R*,*R*)-**101** or (*S*,*S*)-**101**, respectively. Reduction of (*R*,*R*)-**101** with phenylsilane produced (*S*,*S*)-**18** accompanied with small amounts of *meso* compound [36]. Recrystallization of the Rh complex of this mixture afforded enantiomerically pure [Rh((*S*,*S*)-**18**)(COD)]BF_4.

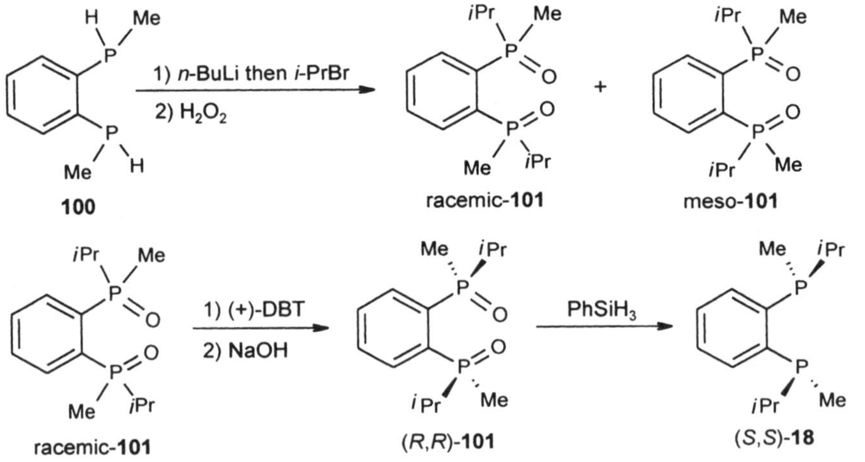

Scheme 19. Example of synthesis of a structurally rigid phosphine ligand

3.4.4
Ferrocene Type Diphosphine Ligand

1,1′-Diphosphine substituted ferrocene ligands have a large P-M-P bite angle at the transition complex, and thus are attractive ligands. P-Chirogenic ferrocene diphosphine **32** was synthesized from 1,1′-dilithioferrocene-TMEDA complex by sequential treatments with *tert*-butyldichlorophosphine, methylmagnesium bromide, and borane-THF complex to afford a mixture of the *dl*-form (50%) and the *meso*-form (10%) in good yield (Scheme 20) [48]. After resolution of the *dl*-form by preparative recycling HPLC, the (*S*,*S*)-form of **104** was reacted with pyrrolidine at 70 °C to furnish the free phosphine ligand (*S*,*S*)-**32** almost quantitatively and complexed with $[Rh(nbd)_2]BF_4$ to produce the desired chiral rhodium cation complex.

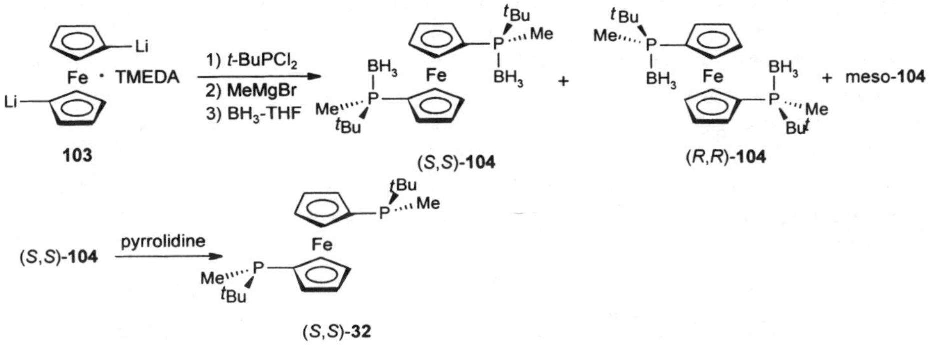

Scheme 20. Example of synthesis of a ferrocenyl-based P-chirogenic phosphine ligand

The wide choice of new methods for the synthesis of P-chirogenic phosphines can be appreciated from the above results,. The synthetic description of this chapter ends here and gives place to the great catalytic potential of these ligands.

4
Asymmetric Transition-Metal Catalyzed Reactions

As mentioned in Sect. 2.3, cleavage of the P-B bond is required prior to complexation with a transition-metal. The following section provides an example for the preparation of two rhodium complexes. Such synthesis, however, may be generalized to almost any other transition-metal/ligand complexes.

4.1
Synthesis of BisP* and MiniPHOS Catalytic Precursors

The phosphorus-boron bond of BisP*-BH_3 **81** is cleaved by treatment with trifluoromethanesulfonic acid in toluene followed by treatment with aqueous KOH or, alternatively, with tetrafluoroboronic acid in dichloromethane and subsequent cleavage with aqueous saturated K_2CO_3, to leave the free phosphine in quantitative yields (Scheme 21) [31, 101]. The absolute configuration of (*S*,*S*)-diphosphines with R=*t*-Bu and R=Cy were determined by X-ray analysis of the corresponding TfOH salt (*t*-Bu-BisP*·2TfOH) and Ru complex (CyBisP*·Ru(C_4H_7)$_2$), respectively. Precatalysts **105** were obtained from the immediate complexation of free phosphines **6** and [Rh(nbd)$_2$]BF_4. X-ray crystallography testified that a single molecule of BisP* binds to the rhodium metal. Although [BisP*Rh(nbd)]BF_4 can be, in some cases, contaminated with substantial amounts of bischelate complexes [Rh(BisP*)$_2$]BF_4, enantiomerically pure [BisP*Rh(nbd)]BF_4 precatalysts are obtained by recrystallization in appropriate solvent. In test asymmetric reactions, the contaminated mixture may be employed directly since both Rh-catalysts contain the same ligand and thus do not affect the optical outcome of the reactions. The following important structural features of the newly designed catalysts were also confirmed: i) the coordinated diphosphines are C_2-symmetric and possess a (*S*,*S*)-configuration; ii) the bulky alkyl groups occupy the *quasi*-equatorial positions and the methyl groups the *quasi*-axial positions, fixing the chelate cycles in a λ-conformation.

The borane group of MiniPHOS was removed using the same protocol (Scheme 21). In contrast, however, reaction of free diphosphines **13** with [Rh(nbd)$_2$]$^+$X$^-$ (X=BF_4, PF_6) afforded in all cases, and independently of the reaction conditions, bischelate complexes **106**. James and Mahajan demonstrated

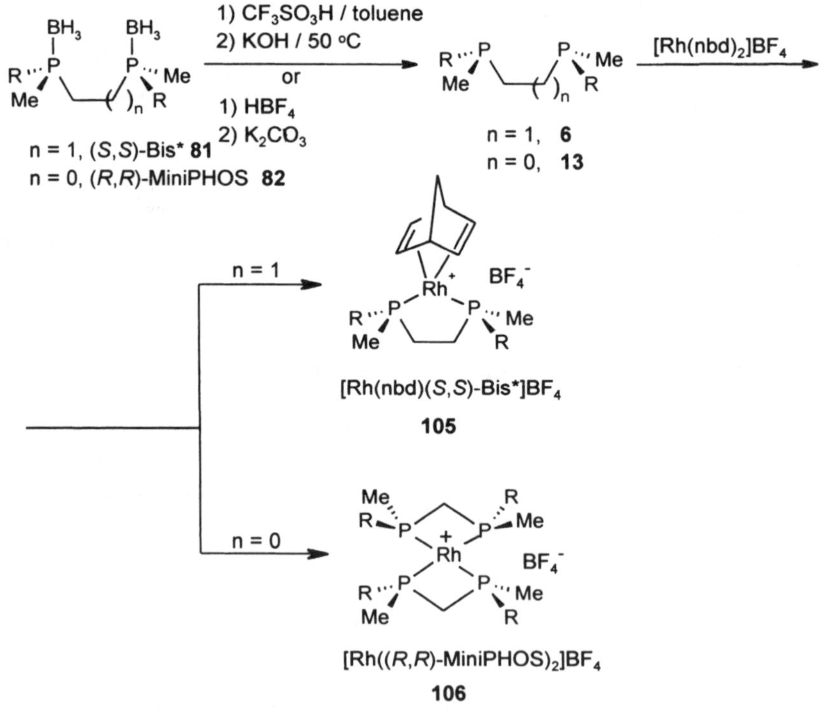

Scheme 21. Example of deboronation of P-stereogenic phosphine-boranes

that, although slower, rhodium-catalyzed asymmetric hydrogenation employing bis(DIOP) species led to higher optical yields than the corresponding monochelate complex [102]. Observation of the X-ray structure of the *t*-Bu-MiniPHOS-rhodium complex showed evidence for (*R*,*R*)-configuration and overall C_2-symmetric environment, where the two bulky groups occupy two diagonal quadrant and the smaller groups the two other quadrant [56]. The MiniPHOS-Rh complex is structurally more rigid than BisP* and forms a highly strained four-membered ring chelate.

4.2 Asymmetric Hydrogenation Reactions

Amongst the several catalytic asymmetric reactions which have been developed to date, enormous progress has been achieved in catalytic enantioselective hydrogenations of the unsaturated bonds [5, 6, 103]. Typically, rhodium(I) and ruthenium(II) complexes, and most often diphosphine-based chiral ligands, have been extensively employed in asymmetric hydrogenation, achieving high reaction rate and excellent enantioselectivity in optimized reaction conditions. This on-going progress constitutes the subject of this section.

4.2.1 *Rh-Catalyzed Asymmetric Hydrogenation Reactions*

4.2.1.1 *Dehydroamino Esters and Itaconic Acid*

Enamide hydrogenations have become a routine test reaction for evaluation of the effectiveness of new chiral ligands [5, 11, 20, 56, 59, 104]. In addition to being a test reaction, it stands as one of the most powerful and economic methods for the production of enantiomerically pure α-amino acid derivatives. Our group [29, 31, 32, 101] and others [35, 42, 43, 45, 47, 52] carried out an extensive series of hydrogenation reactions of dehydroamino acids using the rhodium complexes of P-chirogenic ligands, achieving enantiomeric excesses up to 99.9% (BisP* **6**) [31, 101], 99% (MiniPHOS **13** [29, 101], TangPHOS **17** [35], Jugé's ligand **24** [42] and Nagel's ligand **36**[52]), 92% (Brown's ligand **25**) [43], and 84% (BIPNOR **27**) [45], as shown in Scheme 22.

Conclusions which were drawn from our screening are that the Rh-BisP* **6** and Rh-MiniPHOS **13** are extremely efficient catalysts for the asymmetric hydrogenation of various dehydroamino acids, allowing exceedingly high ees for β-unsubstituted, β-monosubstituted, and β-branched enamines (Scheme 22) [29, 31, 101]. On the other hand, BisP*-Rh catalysts present higher catalytic activity than the corresponding MiniPHOS analogues so that hydrogenation is completed in short reaction time (1–2 h) under moderate H_2 pressure (2 atm) and ambient temperature. However, hydrogenation of β-branched dehydroamino acids requires higher pressure and a longer reaction time. The relatively lower catalytic activity of MiniPHOS can be compensated by carrying out hydrogenation at high pressure, without any loss of optical purity. Interestingly, the enantioselection

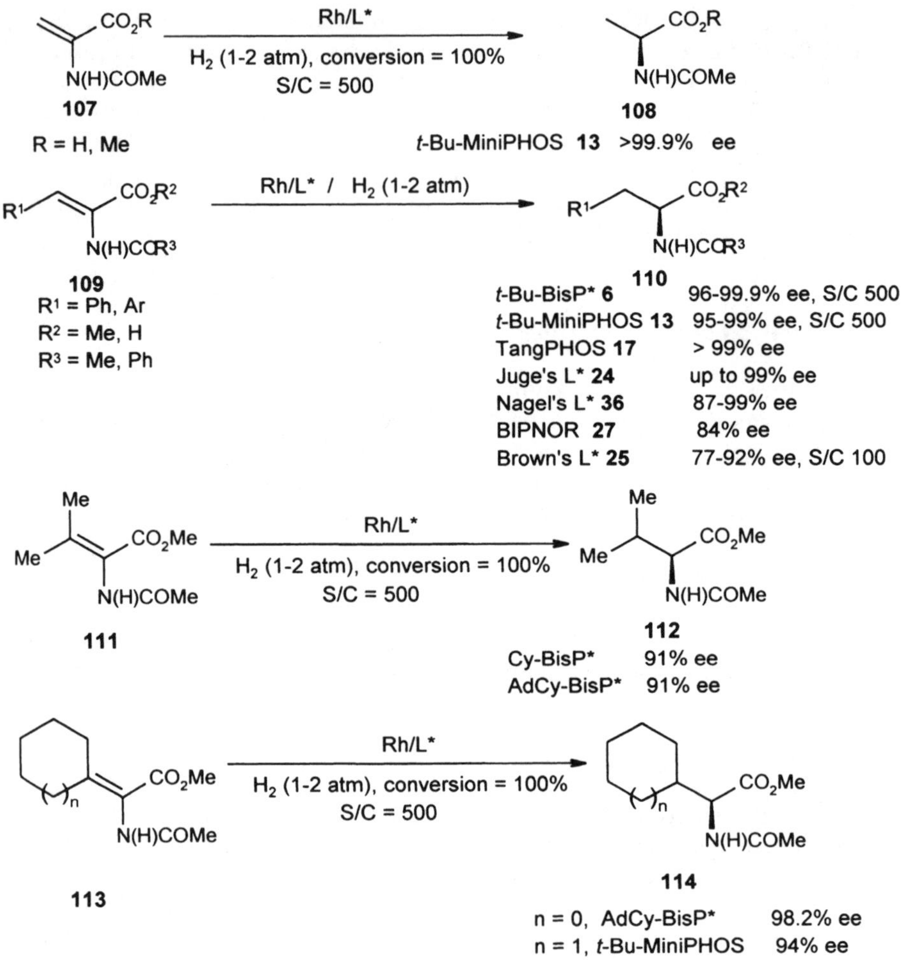

Scheme 22. Examples of Rh-catalyzed asymmetric hydrogenation reactions of dehydroamino esters

power of BisP*-Rh shows a pressure dependence in which the ee decreases as the pressure increases.

Methanol is the solvent usually chosen for the best compromise between enantioselection and reaction yields. It is interesting to note that the BisP*-Rh precatalysts which give the worse results in the asymmetric hydrogenation of β-unsubstituted α-dehydroamino acids **107** or β-monosubstituted ones **109** give the best enantioselectivities for the asymmetric hydrogenation of β-disubstituted α-dehydroamino acids **111** and **113**, and vice versa. Thus *t*-Bu-BisP* and *t*-Bu-MiniPHOS provide optimal results for the hydrogenation of β-unsubstituted and β-monosubstituted dehydroamino acids with almost perfect ees. On the other hand, the Rh-complexes of CyBisP*, AdCyBisP*, and *t*-Bu-MiniPHOS prove the most effective catalysts for the reduction of β-disubstituted dehydroamino acids.

The X-ray crystallographic analysis of the unsymmetrical BisP* shows a strong distortion of the five-membered chelation ring as compared to that of symmetric BisP* [32]. The large difference in the steric repulsions between the bulky substituent borne on one phosphorus atom and the neighboring atoms on the one hand and the other (different) bulky substituent borne on the other phosphorus atom and the same neighboring atoms on the other hand is believed to be responsible for better steric matching with some substrates.

The rigid analogue to BisP*, (*S*,*S*)-1,2-bis(isopropylmethylphosphino)benzene **18**, complexed with rhodium was found to be effective in the asymmetric hydrogenation of the same dehydroamino acid methyl esters affording excellent enantioselectivities (up to 98%) [36]. On the other hand, only modest enantioselection (up to 76% ee) was achieved with the rhodium/flexible ligand **19** complex, suggesting that the conformational flexibility of the seven chelate membered chelate ring favoring conformations deviating from the C_2-symmetry is a detrimental factor to obtaining high enantioselectivity [37].

Although the asymmetric hydrogenation of itaconic acid derivatives is a potential synthetic approach to many useful product [105], lower enantioselectivities are often reported. In contrast with other catalysts, *t*-Bu-BisP*, Ad-BisP*, *t*-Bu-MiniPHOS, BIPNOR **27**, and Brown's ligand **25** gave high to almost perfect ees in the hydrogenation of these substrates (Scheme 23) [101].

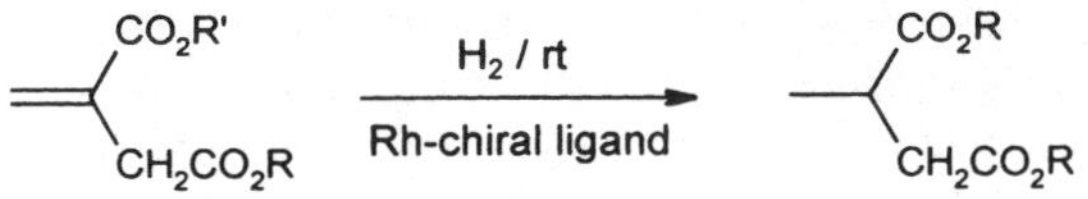

115 R, R' = Me or H

116

t-Bu- or Ad-BisP*,	98.6-99.6% ee
t-Bu-MiniPHOS,	99.9% ee
Brown's L* **25**	80-95% ee
BIPNOR **27**	93% ee

Scheme 23. Rh-catalyzed asymmetric hydrogenation reactions of itaconic acid derivatives

4.2.1.2
Enamides

Several interesting results stemmed from the hydrogenation of various aryl-substituted and alkyl-substituted enamides catalyzed by Rh-BisP* and Rh-MiniPHOS. First, phenylenamides gave the corresponding optically active amides with 96–99% ee (Scheme 24) [106, 107]. However, optical yields of *ortho*-substituted compounds were drastically decreased as compared to the *meta*- or *para*-substituted ones, suggesting that less hindrance around the double bond is favorable for successful asymmetric catalysis. Second, all the studied aryl-substituted enamides gave the *R*-hydrogenated products whereas the enamides bearing bulky alkyl groups such as *tert*-butyl or adamantyl gave the *S*-products in the same reaction conditions (Scheme 24) [106, 107]. This striking difference in the sense of stereoselection was first reported by Burk et al. in the rhodium-catalyzed

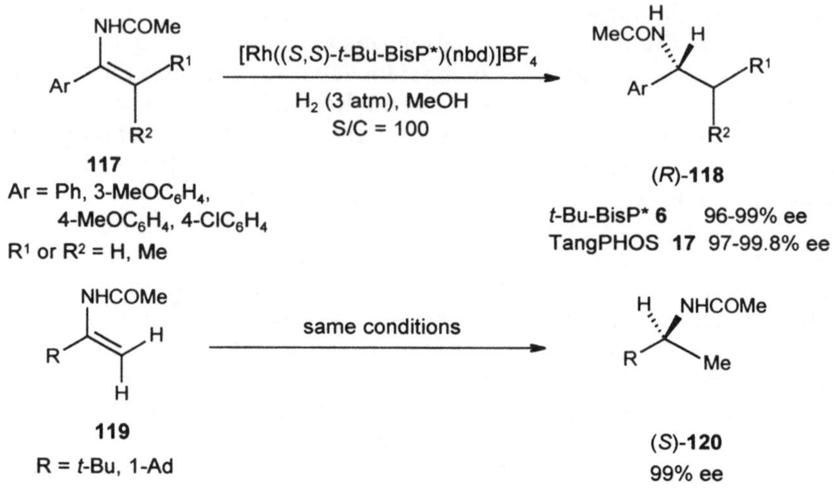

Scheme 24. Examples of Rh-catalyzed asymmetric hydrogenation reactions of enamide derivatives

asymmetric hydrogenation of 1-acetamido-1-phenylethene and 2-acetamido-3,3-dimethyl-1-butene employing (*S*,*S*)-Me-DuPHOS [108, 109]. (*R*)-Absolute configuration was also assigned when the Rh-TangPHOS **17** system was employed for the hydrogenation of α-arylenamides, with exceedingly high TON (10,000) and ees (up to 99.8%). From these results, it seems that the stereochemical environment of the double bond of the enamide considerably affects the sense of the stereoselection. Mechanistic studies based on meticulous NMR studies led to the conclusion that a different mode of coordination of the substrate to the metal is responsible for the opposite sense of stereoselection [106, 107].

The rhodium complex of the (*R*,*R*)-counter-enantiomer of (*S*,*S*)-BisP* achieved a high level of ee (97%) in the asymmetric hydrogenation of 3-methoxy-substituted substrate (*S*)-**122** (Scheme 25), which constitutes a precursor to the acetylcholinesterase inhibitor SDZ-ENA-713 (**123**).

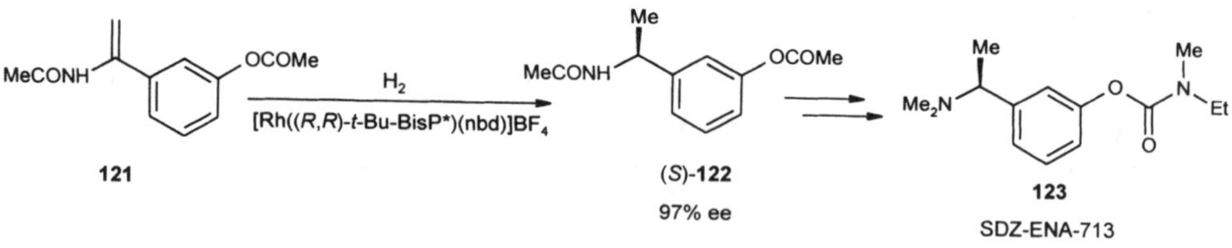

Scheme 25. Rh-catalyzed asymmetric hydrogenation reactions leading to a biologically active compound

4.2.1.3
(*E*)-*β*-(*Acylamino*)*acrylates*

A wide range of optically active natural and unnatural *α*-aminoacids have been synthesized in excellent ees by catalytic asymmetric hydrogenation. On the other hand, the same approach for obtaining *β*-aminoacids has been the subject of less attention [110, 111]. Yet these are key precursors to the synthesis of *β*-lactams [112], *β*-peptides [113, 114], and some medicinally important natural products [116–118]. The rhodium complexes of (*S*,*S*)-BisP* and (*R*,*R*)-MiniPHOS proved to be excellent catalysts for the asymmetric hydrogenation of (*E*)-*β*-(acylamino)acrylates, affording both excellent to almost perfect (*R*)-enantioselectivity and remarkable catalytic efficiency (Scheme 26) [119]. The reaction proceeds smoothly, and is tolerant to both change in solvent and variation in the nature of the substituents (ester and alkyl groups) of the starting substrate. The electron-rich character of BisP* and MiniPHOS results in an increased affinity of their rhodium complex to dihydrogen and thus faster hydrogenation as compared to, e.g., (*R*,*R*)-Me-DuPHOS [111].

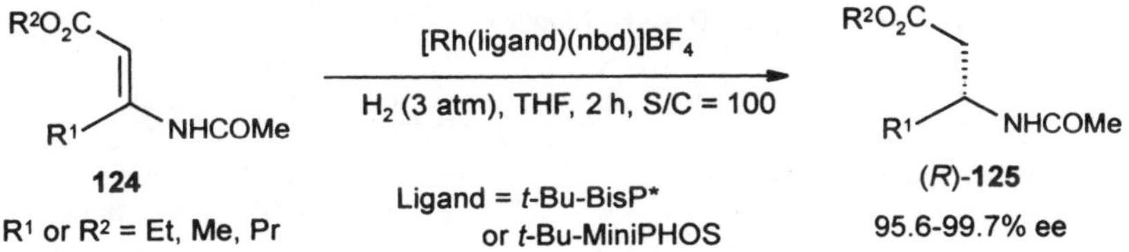

Scheme 26. Rh-catalyzed asymmetric hydrogenation reactions of (*E*)-*β*-(acylamino)acrylates

4.2.1.4
Ethenephosphonate

The Rh(I)-catalyzed asymmetric hydrogenation of dimethyl 1-benzoyloxyethenephosphonate 2 using *t*-Bu-BisP* as the chiral ligand gave the corresponding (*S*)-product in 88% ee (Scheme 27) [120], enantioselectivity being comparable to those observed by Burk et al. [121].

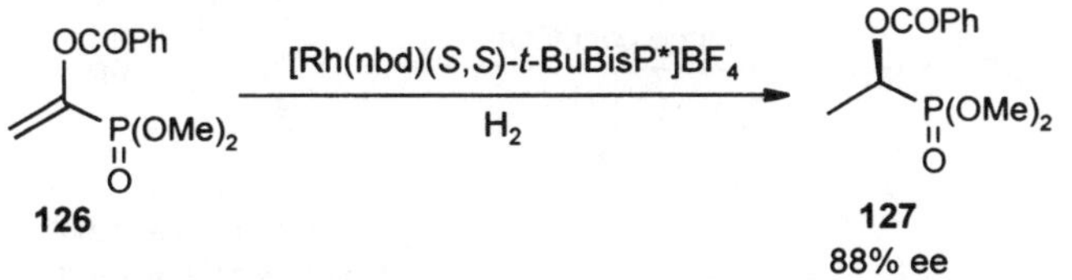

Scheme 27. Example of the Rh-catalyzed asymmetric hydrogenation reaction of ethenephosphonate

4.2.2
Ru-Catalyzed Asymmetric Hydrogenation of β-Ketoesters

The broad applicability of BisP* complexed to a transition-metal was further exemplified by the ruthenium-catalyzed asymmetric hydrogenation of β-keto esters [122]. The BisP*-$RuBr_2$ precatalyst can be readily prepared by reaction between BisP* and (COD)Ru(2-methylallyl)$_2$, followed by treatment with methanolic HBr. Several kinds of keto esters were subjected to asymmetric hydrogenations under mild conditions (6 atm H_2, 50–70 °C), yet only β-keto esters were smoothly hydrogenated, with high chemical yields and enantioselectivities up to 98% (Scheme 28). The general trend was that the best ees were obtained with *t*-Bu-BisP*. β-Keto amides and β-keto phosphonates provide also high enantioselectivities. Furthermore, the Ru-BIPNOR complex permitted ees between 74 and 95% in the hydrogenation of acetophenone and acetonaphthanone [45].

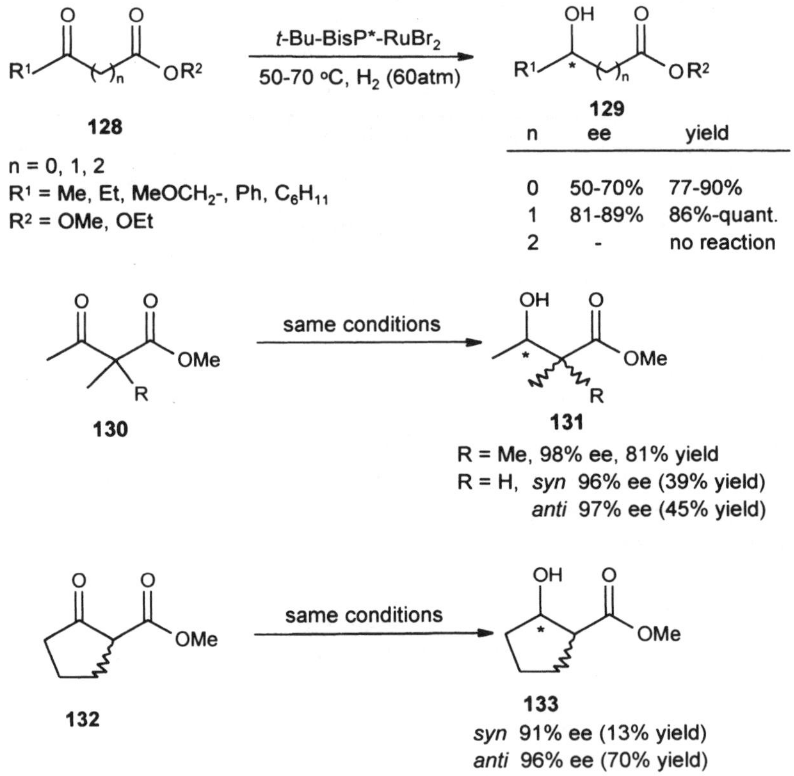

Scheme 28. Examples of Ru-catalyzed asymmetric hydrogenation reactions of β-keto esters

4.3
Rh-Catalyzed Asymmetric Hydrosilylation

The rhodium complex of P-chirogenic diphosphine ferrocene (*S*,*S*)-**32** enabled one to obtain the highest selectivity in the rhodium-catalyzed asymmetric hydrosilylation of ketones amongst the known *cis*-chelating diphosphine of that time [48, 123]. Other *cis*-diphosphine ligands only led to low-to-moderate ees, whereas *trans*-diphosphine ligands were known to exhibit high selectivity [124]. Thus asymmetric hydrosilylation of substrates such as acetophenone, substituted phenyl methyl ketones, naphthyl methyl ketones, and phenethyl methyl ketones proceeded in the presence of a 100:1:150 ratio of ketone/Rh catalyst/naphthylphenylsilane to afford the (*S*)-alcohol in excellent yields (Scheme 29). Greater selectivity was achieved by the use of 1-naphthylphenylsilane as opposed to diphenylsilane (probably due to greater steric hindrance), and by the use of aryl-substituted methyl ketones as opposed to alkyl-substituted ones. In such cases, ees up to 92% were demonstrated [48]. Similarly to the rhodium-catalyzed asymmetric hydrogenation of enamides using Rh-BisP* mentioned previously [107], *ortho*-substituted tolyl methyl ketone gives low enantiomeric excess (44% ee) as compared to the corresponding *meta*- and *para*-substituted derivatives (84–85% ee).

Similar asymmetric hydrosilylation reactions were also performed using Rh-(*R*,*R*)-*t*-Bu-MiniPHOS, and the enantioselectivities obtained (80–97% ee) [29] are comparable with those obtained by use of the most effective ligands [125].

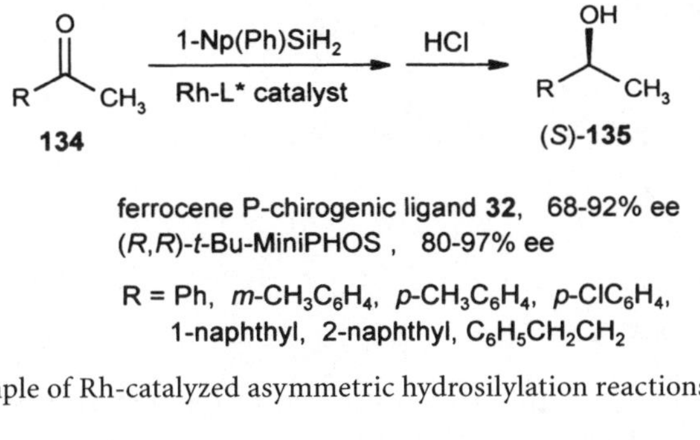

ferrocene P-chirogenic ligand **32**, 68-92% ee
(*R*,*R*)-*t*-Bu-MiniPHOS , 80-97% ee

R = Ph, *m*-$CH_3C_6H_4$, *p*-$CH_3C_6H_4$, *p*-ClC_6H_4, 1-naphthyl, 2-naphthyl, $C_6H_5CH_2CH_2$

Scheme 29. Example of Rh-catalyzed asymmetric hydrosilylation reactions

4.4
Pd-Catalyzed Asymmetric Allylic Alkylation

Pd-catalyzed asymmetric allylic alkylation is a typical catalytic carbon-carbon bond forming reaction [126–128]. The Pd-complex of the ligand (*R*)-**3b** bearing methyl, 2-biphenyl and cyclohexyl groups as the three substituents attached to the P-chirogenic phosphorus atom was found to be in situ an efficient catalyst in the asymmetric allylic alkylation of 1-acetoxy-1,3-diphenylprop-2-en (**4**) with malonate derivatives in the presence of *N*,*O*-bis(trimethylsilyl)acetamide (BSA) and potassium acetate, affording enantioselectivity up to 96% and quantitative yields (Scheme 30) [41]. The reaction was revealed to be merely affected by the change in the nature of the additive, but performed best in dichloromethane as the solvent.

The Pd-complexes of P-chirogenic phosphine/sulfide hybrid ligands **14a,b** were also employed in situ and tested in the same reactions. Ligand **14** possessing the *t*-Bu group on the phosphine atom and the phenyl substituent on the sulfur atom was revealed to produce the best compromise between chemical and optical yields, achieving the (*R*)-adduct in almost quantitative yields and high enantioselectivity up to 90%. Interestingly, ligand **14b**, in which the two donor atoms are linked by an ethylene bridge, as opposed to the methylene chain in **14a**, gave adducts of opposite configuration at the carbon center, likely because of a different orientation of the phenyl group on the sulfur atom in the five-membered chelate ring structure [10].

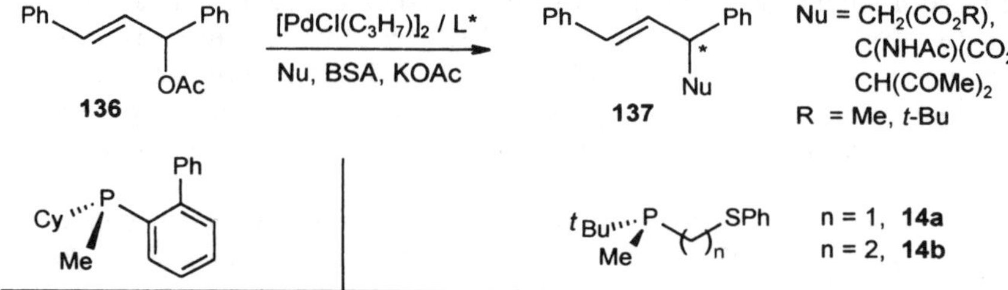

(*R*)-**23b**
79-96% ee, (*S*)-product
quant. yield

n	ee %	config. product	yield (%)
1	85-90	(*R*)	94-99
2	53-74	(*S*)	50-99

Scheme 30. Example of Pd-catalyzed asymmetric allylic alkylation

4.5 Michael-Type Reactions

The enantioselective 1,4-addition addition of organometallic reagents to α,β-unsaturated carbonyl compounds, the so-called Michael reaction, provides a powerful method for the synthesis of optically active compounds by carbon-carbon bond formation [129]. Therefore, symmetrical and unsymmetrical MiniPHOS phosphines were used for in situ preparation of copper-catalysts, and employed in an optimization study on Cu(I)-catalyzed Michael reactions of diethylzinc to α,β-unsaturated ketones (Scheme 31) [29, 30]. In most cases, complete conversion and good enantioselectivity were obtained and no 1,2-addition product was detected, showing complete regioselectivity. Of interest, the enantioselectivity observed using Cu(I) *directly* in place of Cu(II) allowed enhanced enantioselectivity, implying that the chiral environment of the Cu(I) complex produced by in situ reduction of Cu(II) may be less selective than the one with preformed Cu(I).

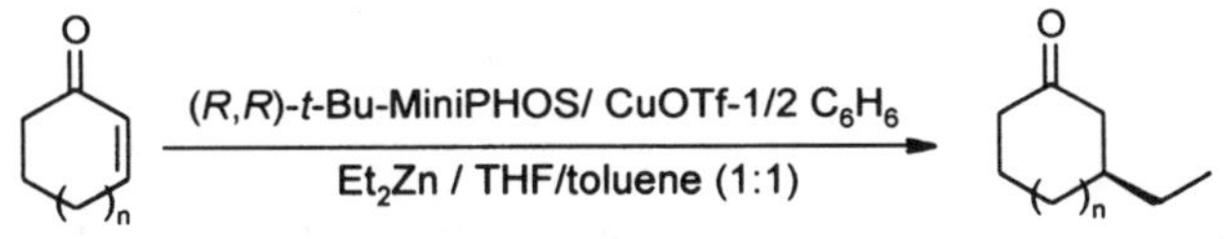

n = 0 - 2
138

n = 0, **139**
n = 1, **140**
n = 2, **141**

n	ee (%)	yield (%)
0	43	71
1	85	92
2	75	92

Scheme 31. Example of Cu-catalyzed asymmetric Michael addition

The best conditions defined permitted isolation of (*S*)-3-ethylcyclopentan-1-one (**139**) in 43% ee (71% yield), (*S*)-3-ethylcyclohexan-1-one (**140**) with 85% ee and high yield, and (*S*)-3-ethylcycloheptan-1-one (**141**) in 75% ee (92% yield) [30]. These results were satisfactory given the fact that the same enantioselective Michael addition using other chiral disphosphine ligands such as BINAP or DuPHOS is known to have so far proceeded in low enantioselectivity [130]. Moreover, since until very recently [131], the catalytic conjugate addition of alkylmetals to a five-membered ring has been reported as being significantly less efficient and selective than reactions of larger rings [132–136].

5
Conclusion

The considerable number of P-chirogenic ligands which have been synthesized in the past ten years and described in this chapter testifies to the increasing importance of the catalysts bearing such phosphines in the field of asymmetric catalysis. Not only have been designed novel ligands shaped for excellent enantioselection but also new methodologies have been elaborated to make their synthesis economically valuable and comfortable. The direct contact between the chiral center and the transition-metal (i.e., the reaction site), achieving exceedingly high levels of enantioselectivity, along with the great affinity of hydrogen for electron-rich metal catalysts bearing trialkylated P-chirogenic phosphine ligands, enabling attractive turnover numbers, are considered to be the main advantages in using ligand bearing the chiral center at a phosphorus atom. Due to the sudden interest that synthetic chemists have shown in this particular type of phosphine ligands, research focus has been devoted to a large extent to the synthetic part. Thus the full potential of these chiral catalysts has not been explored yet, and it is expected that exceptionally high enantio-, diastereo-, regio-, and chemoselectivities may be obtained in future, placing the P-chirogenic phosphine ligands amongst the leading ligands for the preparation of enantiomerically pure compounds.

6
References

1. Whitesell JK (1989) Chem Rev 1581
2. Brunner H (1988) In: Eliel EL, Wilen SH (eds) Topics in stereochemistry, vol 18. Wiley, New York, p 129
3. Hayashi T (2000) Acc Chem Res 33:354
4. Lagasse F, Kagan HB (2000) Chem Pharm Bull 48:315
5. Ojima I (ed) (2000) Catalysis asymmetric synthesis. Wiley-VCH, Weinheim
6. Noyori R (1994) Asymmetric catalysis in organic synthesis. Wiley, New York
7. Enders D, Peters R, Runsink J, Bats JW (1999) Org Lett 1:1863
8. Evans DA, Campos KR, Tedrow JS, Michael EF, Gagné MR (2000) J Am Chem Soc 122:7905
9. Nakano H, Okuyama Y, Yanagida M, Hongo H (2001) J Org Chem 66:620
10. Sugama H, Saito H, Danjo H, Imamoto T (2001) Synthesis 15:2348
11. Kagan HB (1985) Chiral ligands for asymmetric catalysis. In: Morrison JD (ed) Asymmetric synthesis, vol 5. Academic Press, New York, chap 1
12. Mann G, Incarvito C, Rheingold AL, Harrtwig JF (1999) J Am Chem Soc 121:3224
13. Wolfe JP, Simger RA, Yang BH, Buchwald SL (1999) J Am Chem Soc 121:9550
14. Littke AF, Dai C, Fu GC (2000) J Am Chem Soc 122:4020
15. Torraca KE, Kuwabe S, Buchwald SL (2000) J Am Chem Soc 122:12,907
16. Wolfe JP, Tomori H, Sadighi JP, Yin J, Buchwald SL (2000) J Org Chem 65:1158
17. Ueda M, Miyaura N (2000) J Org Chem 65:4450
18. Brown JM, Chalnoer PA (1983) In: Pignolet NH (ed) Homogeneous catalysis with metal phosphine complexes. Plenum, New York, chap 4
19. Bosnich B (ed) (1986) Asymmetric catalysis. Martinus Nijhoff Publishers, Dordrecht
20. Koenig KE (1985) Applicability of asymmetric hydrogenation. In: Morrison JD (ed) Asymmetric synthesis, vol 5. Academic Press, New York, chap 3
21. Yamanoi Y, Imamoto T (1999) Rev Heteroatom Chem 20:227
22. Horner L, Siegel H, Buthe H (1968) Angew Chem Int Ed Engl 7:942
23. Knowles WS, Sabacky MJ (1968) J Chem Soc Chem Commun 1445
24. Knowles WS, Sabacky MJ, Vineyard BD (1972) J Chem Soc Chem Commun 10
25. Knowles WS, Sabacky MJ, Vineyard BD, Weinkauff DJ (1975) J Am Chem Soc 97:2567
26. Imamoto T, Tsuruta H, Wada Y, Masuda H, Yamaguchi K (1995) Tetrahedron Lett 36:8271
27. Muci AR, Campos KR, Evans DA (1995) J Am Chem Soc 117:9075
28. Airey AL, Swiegers GF, Willis AC, Wild SB (1995) J Chem Soc, Chem Commun 693
29. Yamanoi Y, Imamoto T (1999) J Org Chem 64:2988
30. Taira S, Crépy KVL, Imamoto T (2002) Chirality 14:386
31. Imamoto T, Watanabe J, Wada Y, Masuda H, Yamada H, Tsuruta H, Matsukawa S, Yamaguchi K (1998) J Am Chem Soc 120:1635
32. Ohashi A, Imamoto T (2001) Org Lett 3:373
33. Kaloun EB, Merdès R, Genêt J-P, Uziel J, Jugé S (1997) J Organometal Chem 529:455
34. Yang H, Alvarez M, Lugan N, Mathieu R (1995) J Chem Soc, Chem Commun 1721
35. Tang W, Zhang X (2002) Angew Chem 114:1682
36. Miura T, Imamoto T (1999) Tetrahedron Lett 40:4833
37. Gridnev ID, Higashi N, Imamoto T (2001) Organometallics 20:4542
38. Liu A, Mok KF, Leung P-H (1997) Chem Commun 2397
39. Brenchley G, Fedouloff M, Mahom MF, Molloy KC, Wills M (1995) Tetrahedron 51: 10,581
40. Leung P-H, Lang H, White AJP, Williams D (1998) Tetrahedron:Asymmetry 9:2961
41. Tsuruta H, Imamoto T (2001) Synlett 999
42. Moulin D, Darcel C, Jugé S (1999) Tetrahedron: Asymmetry 10:4729
43. Carmichael D, Doucet H, Brown JM (1999) Chem Commun 261
44. Wolfe B, Livinghouse T (1998) J Am Chem Soc 120:5116
45. Robin F, Mercier F, Ricard L, Mathey F, Spagnol M (1997) Chem Eur J 3:1365
46. Hamada T, Buchwald SL (2002) Organic Lett 4:999

47. Nettekoven U, Widhalm M, Kamer PCJ, Van Leeuwen PWNM (1997) Tetrahedron: Asymmetry 8:3185
48. Tsuruta H, Imamoto T (1999) Tetrahedron: Asymmetry 10:877
49. Bader A, Pabel M, Wild SB (1994) J Chem Soc Chem Commun 1405
50. Hamada Y, Matsuura F, Oku M, Hatano K, Shioiri T (1997) Tetrahedron Lett 38:8961
51. Dahlenburg L, Kurth V (1998) Eur J Inorg Chem 597
52. Nagel U, Krink T (1993) Angew Chem Int Ed Engl 32:1052
53. Yamago S, Yanagawa M, Nakamura E (1994) J Chem Soc, Chem Commun 2093
54. Dang TP, Kagan HB (1971) J Chem Soc Chem Commun 481
55. Vineyard BD, Knowels WS, Sabacky GL, Bachman GL, Weinkauff DJ (1977) J Am Chem Soc 99:5946
56. Knowles WS (1983) Acc Chem Res 16:106
57. Pietrusiewicz KM, Zablocka M (1994) Chem Rev 94:1375
58. Valentine DJ (1984) Preparation of the enantiomers of compounds containing chiral phosphorus centers. In: Morrison JD, Scott JW (eds) Asymmetric synthesis, vol 4. Academic Press, New York, chap 3
59. Burk MJ, Feaster JE, Nugent WA, Harlow RL (1993) J Am Chem Soc 115:10,125
60. Burk MJ (2000) Acc Chem Res 33:363
61. Korpium O, Lewis RA, Chikos J, Mislow K (1968) J Am Chem Soc 90:4842
62. Lewis RA, Mislow K (1969) J Am Chem Soc 91:7009
63. Lewis RA, Naumann K, Debriun KE, Mislow K (1969) J Chem Soc Chem Commun: 1010
64. Hoffmann H, Schellenbeck P (1966) Chem Ber 99:1134
65. Kosolapoff GM, Brown ADJ (1967) J Chem Soc 1789
66. Brown ADJ, Kosolapoff GM (1968) J Chem Soc 839
67. Imamoto T, Sato K, Johnson CR (1985) Tetrahedron Lett 26:783
68. Horner L, Balzer WD (1965) Tetrahedron Lett 6:1157
69. Naumann K, Zon G, Mislow K (1969) J Am Chem Soc 91:7012
70. Marsi KL (1974) J Org Chem 39:265
71. Coumbe T, Lawrence NJ, Muhammad F (1994) Tetrahedron Lett 35:625
72. Henson PD, Naumann K, Mislow K (1969) J Am Chem Soc 91:5645
73. Imamoto T, Oshiki T, Onozawa T, Kusumoto T, Sato K (1990) J Am Chem Soc 112:5244
74. Imamoto T, Kikuchi S, Miura T, Wada Y (2001) Organic Lett 3:87
75. Ohff M, Holz J, Quirmbach M, Borner A (1998) Synthesis 1391
76. Carboni B, Monnier L (1999) Tetrahedron 55:1197
77. Imamoto T (1993) Pure Appl Chem 65:655
78. Imamoto T, Kusumoto T, Suzuki N, Sato K (1985) J Am Chem Soc 107:5301
79. Brisset H, Gourdel Y, Pellon P, Le Corre M (1993) Tetrahedron Lett 34:4523
80. McKinstry L, Livinghouse T (1994) Tetrahedron Lett 35:9319
81. McKinstry L, Livinghouse T (1994) Tetrahedron 50:6145
82. Nudelman A, Cram DJ (1968) J Am Chem Soc 90:3869
83. Korpium O, Mislow K (1967) J Am Chem Soc 89:4784
84. Imamoto T, Tsuruta H (1996) Chemistry Lett 707
85. Oshiki T, Hikosaka T, Imamoto T (1991) Tetrahedron Lett 32:3371
86. Koide Y, Sakamoto A, Imamoto T (1991) Tetrahedron Lett 32:3375
87. Imamoto T, Matsuo M, Nonomura T, Kishikawa K, Yanagawa M (1993) Heteroat Chem 4:475
88. Oshiki T, Imamoto T (1992) J Am Chem Soc 114:3975
89. Schmidbaur H (1980) J Organomet Chem 200:287
90. Maryanoff CA, Maryanoff BE, Tang R, Mislow K (1973) J Am Chem Soc 5839
91. Jones AB (1995) In: Paquet LA (ed) Encyclopedia of reagents for organic synthesis, vol 6. Wiley, New York, p 3832
92. Fuhrhop J, Penzlin G (1994) In: Organic synthesis. VCH, Winheim
93. Nagata K, Matsukawa S, Imamoto T (2000) J Org Chem 65:4185
94. Ohashi A, Imamoto T (2001) Tetrahedron Lett 42:1099
95. Miura T, Yamada H, Kikuchi S, Imamoto T (2000) J Org Chem 65:1877
96. Baechler RD, Mislow K (1970) J Am Chem Soc 92:3090

97. Green G, Griffith WP, Hollinshead DM, Ley SV, Schröder M (1984) J Chem Soc Perkin Trans 1:681
98. Imamoto T, Hirose K, Amano H (1996) Main Group Chem. 1:331
99. Matsukawa S, Sugama H, Imamoto T (2000) Tetrahedron Lett 41:6461
100. Kyba EP, Liu S-T, Harris RL (1983) Organometallics 2:1877
101. Gridnev ID, Yamanoi Y, Higashi N, Tsuruta H, Yasutake M, Imamoto T (2001) Adv Synth Catal 343:118
102. James BR, Mahajan D (1985) J Organomet Chem 279:31
103. Brown JB (1999) Hydrogenation of functionalised carbon-carbon double bonds. In: Jacobsen EN, Pfaltz A, Yamamoto H (eds) Comprehensive asymmetric catalysis, vol 1. Springer, Berlin Heidelberg New York, chap 5.1
104. Noyori R, Takaya H (1990) Acc Chem Res 23:345
105. Burk MJ, Bienewald F, Harris M, Zanotti-Gerosa A (1998) Angew Chem Int Ed Engl 37:1931
106. Gridnev ID, Higashi N, Imamoto T (2000) J Am Chem Soc 122:10,486
107. Gridnev ID, Yasutake M, Higashi N, Imamoto T (2001) J Am Chem Soc 123:5268
108. Burk MJ, Wang YM, Lee JR (1996) J Am Chem Soc 118:5142
109. Burk MJ, Casy G, Johnson NB (1998) J Org Chem 63:6084
110. Lubell WD, Kitamura M, Noyori R (1991) Tetrahedron: Asymmetry 2:543
111. Zhu G, Chen Z, Zhang X (1999) J Org Chem 64:6907
112. E. Juaristi (ed) (1997) Enantioselective synthesis of β-amino acids. Wiley-VCH, New York
113. Gellman SH (1998) Acc Chem Res 31:173
114. Cole DC (1994) Tetrahedron 50:9517
115. Hart DJ, Ha D-C (1989) Chem Rev 89:1447
116. Deng L, Jacobsen EN (1992) J Org Chem 57:4320
117. Wang Z-M, Kolb HC, Sharpless KB (1994) J Org Chem 59:5104
118. Guenard D, Voegelein FG, Potier P (1993) Acc Chem Res 115:1151
119. Yasutake M, Gridnev ID, Higashi N, Imamoto T (2001) Organic Lett 3:1701
120. Gridnev ID, Higashi N, Imamoto T (2001) J Am Chem Soc 123:4631
121. Burk MJ, Stammers TA, Straub JA (1999) Org Lett 1:387
122. Yamano T, Taya N, Kawada M, Huang T, Imamoto T (1999) Tetrahedron Lett 40:2577
123. Brunner H, Nishiyama H, Itoh K (1993) Asymmetric hydrosilylation. In: Ojima I (ed) Catalytic asymmetric synthesis. Wiley-VCH, New York, chap 6
124. Sawamura M, Kuwano R, Ito Y (1994) Angew Chem, Int Ed Engl 33:111
125. Kuwano R, Uemura T, Saitoh M, Ito Y (1999) Tetrahedron Lett 40:1327
126. Hayashi T (1993) Asymmetric allylic substitution and grignard cross-coupling. In: Ojima I (ed) Catalytic asymmetric synthesis. Wiley-VCH, New York, chap 7–1
127. Trost BM, Vranken DLV (1996) Chem Rev 96:395
128. Consiglio G, Waymouth RM (1989) Chem Rev 89:257
129. Tomioka K, Nagaoka Y (1999) Conjugate addition of organometallic reagents. In: Jacobsen EN, Pfaltz A, Yamamoto H (eds) Comprehensive asymmetric catalysis, vol 3. Springer, Berlin Heidelberg New York, chap 31.1
130. Alexakis A, Burton J, Vastra J, Mangeney P (1997) Tetrahedron: Asymmetry 8:3987
131. Degrado SJ, Mizutani H, Hoveyda AH (2001) J Am Chem Soc 123:755
132. Feringa BL, Pineschi M, Arnold LA, Imbos R, de Vries AHM (1997) Angew Chem Int Ed Engl 36:2620
133. Naasz R, Arnold LA, Pineschi M, Keller E, Feringa BL (1999) J Am Chem Soc 121:1104
134. Hu X, Chen H, Zhang X (1999) Angew Chem Int Ed Engl 38:3518
135. Alexakis A, Benhaim C, Fournioux X, Heuvel AVD, Levèque J-M, March S, Rosset S (1999) Synlett 1811
136. Yan MZZ-Y, Chan ASC (2000) Chem Commun 115

Top Curr Chem (2003) 229: 41–73
DOI 10.1007/b11149

New Trends in Ylide Chemistry

Marc Taillefer[1] · Henri-Jean Cristau[2]

Laboratoire de Chimie Organique (U.M.R. 5076), Ecole Nationale Supérieure de Chimie, 8, rue de l'Ecole Normale, 34296 Montpellier Cedex 5, France
[1] *E-mail: taillefe@cit.enscm*
[2] *E-mail: cristau@cit.enscm.fr*

This review concerns in the first part the works published during the last three years on the synthesis and reactivity of stabilized ylides *C*-substituted by electron-withdrawing groups (COR, CO_2R, CN, etc.). The second part deals with the works published in the same period on the chemistry of phosphorus ylides mainly *C*-substituted by heteroatoms of groups 1–16 (metals, metalloids and nonmetal elements: Li, Ba, Ca, Ti, Zr, Nb, Mo, Re, Fe, Ru, Rh, Pd, Pt, Au, Zn, Hg, B, Si, Sn, N, P, As, Sb, O, S, Te).

Keywords. Phosphonium, Ylide, Diylide, Wittig, Coordination, Reactivity

1 Introduction

Disregarding the still tremendous applications in the Wittig carbonyl olefination, the latest trends in the chemistry of phosphonium ylides, including exclusively P-C bonds, concerns mainly the chemistry of cumulene ylides, the preparation and reactivity of stabilized ylides, and also the chemistry of phosphorus ylides C-substituted by metal, metalloids, and nonmetals. We have reviewed the last two fields for a period covering the works published in the last three years, thus updating a part of the book of Oleg I. Kolodiazhnyi "Phosphorus ylides, chemistry and application in organic synthesis", published in 1999 [29]. Our first section is mainly dedicated to the preparation and reactivity of stabilized ylides *C*-substituted by an electron-withdrawing group (COR, CO_2R, CN, etc.). The second – which concerns the chemistry of phosphorus ylides *C*-substituted by heteroatoms – is extremely varied. Indeed almost all the groups of the periodic table are represented in the *C*-substitution (Li, Ba, Ca, Ti, Zr, Nb, Mo, Re, Fe, Ru, Rh, Pd, Pt, Au, Zn, Hg, B, Si, Sn, N, P, As, Sb, O, S, Te) thus leading to a variety of different properties and applications for the corresponding ylides.

2 Phosphonium Ylides: Preparation and Reactivity

2.1 Preparation of Stabilized Ylides

2.1.1 *Preparation by Addition of Tertiary Phosphines to Acetylenic Compounds Substituted by Electron-Withdrawing Groups*

Triphenylphosphine gives Michael additions to the activated triple bond of acetylene dicarboxylic esters in presence of acidic compounds HY (Scheme 1). The reactions take place easily at room temperature, even at –10 °C [1], through formation of intermediate activated vinylic phosphonium salts, which undergo a subsequent Michael addition of HY. The reactions afford various stabilized ylides which can be isolated in high yields or undergo possibly evolution, for example by intramolecular Wittig reaction [2].

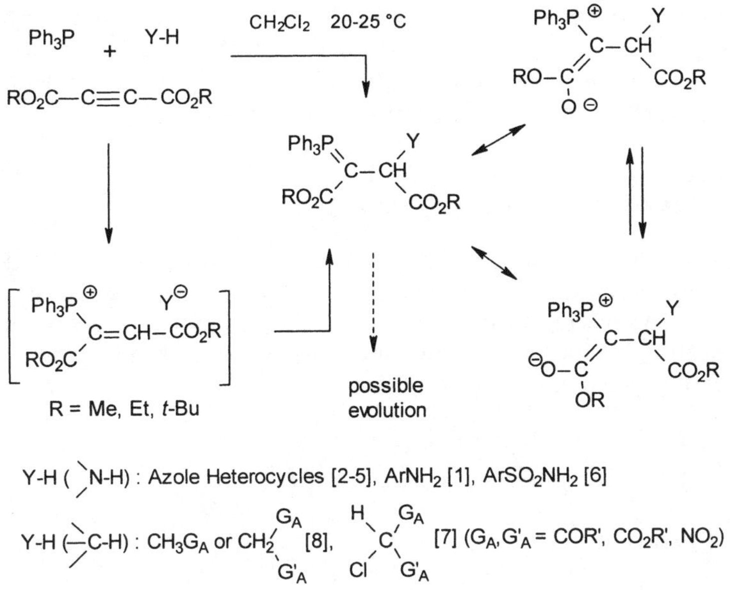

Scheme 1

The acidic compounds used can be compounds with N-H bonds (aromatic primary amines [1], azole heterocycles [2–5], sulfonamides [6]), or enolizable compounds with activated C-H bonds [7, 8].

A related preparation of specific stabilized phosphonium ylides corresponds to the reaction of triarylphosphines with acetylene dicarboxylic esters in presence of fullerene, which affords a cyclopropanyl-fullerene substituted stabilized phosphonium ylide [9] or the corresponding evolution products [10].

2.1.2
Preparation by Addition of Tertiary Phosphine Oxides to Acetylene Dicarboxylic Esters

The reaction takes place probably by a kind of "inverse Wittig reaction", corresponding to the thermal dissociation of an oxaphosphetene resulting from a [2+2] cycloaddition between the phosphine oxide and the activated acetylenic compounds (Scheme 2) [11, 12].

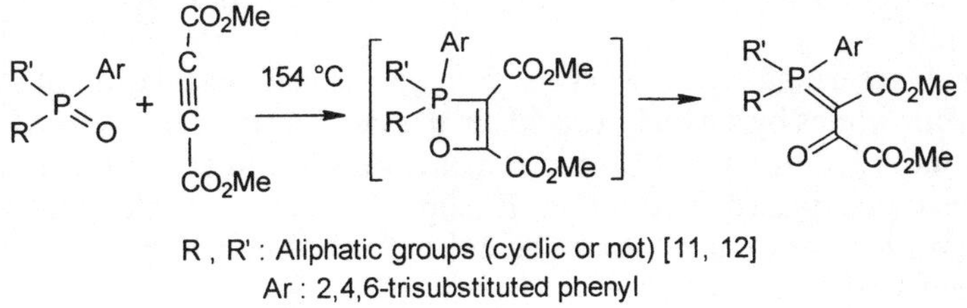

Scheme 2

The reaction affords in variable yields (19–85%) a stabilized phosphonium ylide exhibiting two electron withdrawing substituents on the ylidic carbon α to the phosphorus atom, and requires apparently a dialkylarylphosphine oxide incorporating an overcrowded 2,4,6-trisubstituted aryl group.

2.1.3
Preparation by Modification of the Side Chain

Among various modifications of the side chain of stabilized ylides [13, 14] can be pointed out the preparation and transformation of the phenyliodonio α-substituted phosphonium ylides (Scheme 3) [15]. These compounds represent a potentially useful class of reagents, in which the iodonium group can be further substituted by nucleophiles such as PhSLi.

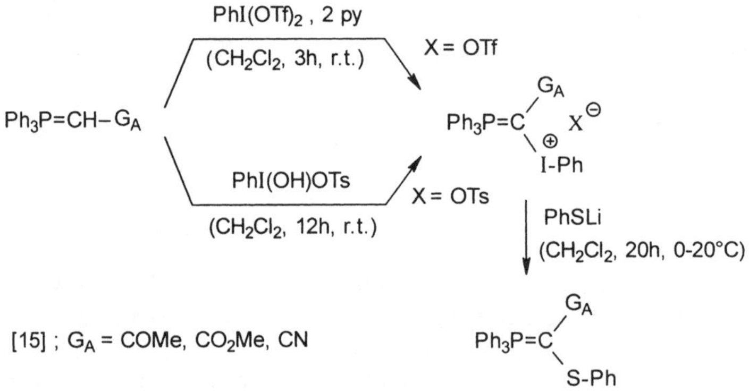

Scheme 3

2.1.4
Preparation by Acylation of Phosphonium Ylides

Since the pioneering work of H. J. Bestmann and coworkers in the 1960s, acylation of phosphonium ylides is a well known process for the preparation of β-oxophosphonium ylides. The classical way (a) using acylating agents such as acyl chlorides and in situ transylidation is still useful (Scheme 4) [16, 17].

However, an interesting variant (way b) avoiding the loss of one equivalent of the starting ylide as a consequence of the transylidation reaction was recently developed using a zinc promoted acylation of stabilized ylides by acyl chlorides [18].

Further, Wasserman and coworkers developed a direct acylation of stabilized phosphonium ylides by carboxylic acids in presence of the EDCI/DMAP (way c). This last method allows the introduction of α-aminoacid structures into the resulting β-oxo phosphorus ylides [19–25], opening the way to the total synthesis of depsipeptide elastase inhibitors [22, 24] or cyclic peptidic protease inhibitor Eurystatin A [20].

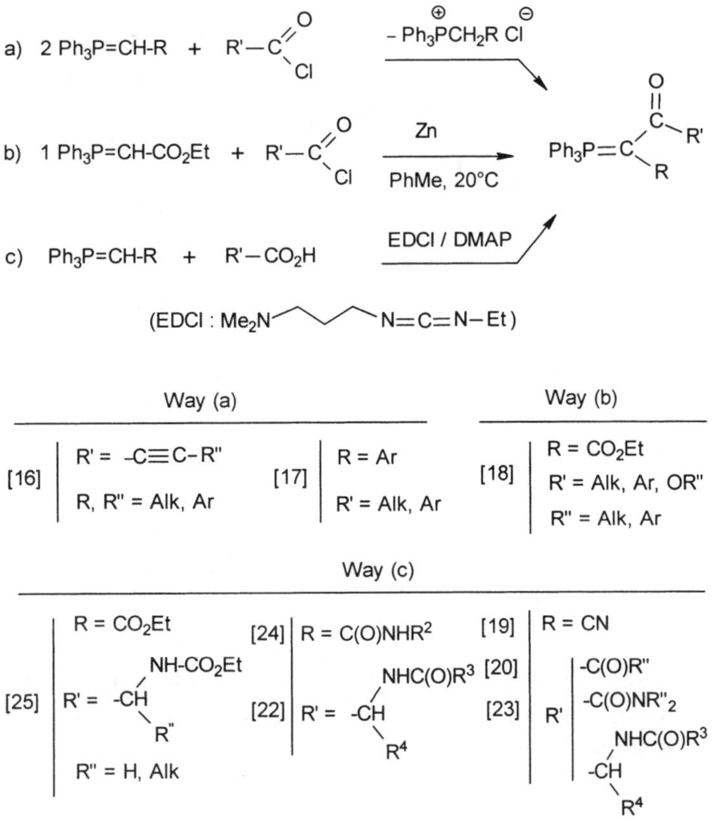

Scheme 4

2.2
Reactivity of Stabilized Ylides

2.2.1
Thermolysis of β-Oxo Phosphorus Ylides

Savignac et al. [26] excellently reviewed this topic, but Aitken and coworkers investigated further the flash vacuum pyrolysis (FVP) of the β-oxophosphorus ylides (Scheme 5).

The thermolysis of various substituted phosphonium ylides between 600 °C and 900 °C can afford either substituted alkynes [16, 25, 27] or cyclic dienes [20] by extrusion of Ph_3PO, or new stabilized ylides by cyclization of the functional groups [27, 28].

Particularly interesting are the results obtained with the phosphonium ylides including an acyl rest derived from aminoacid: if the N-H bond reactivity is blocked by an amide protection, the alkyne formation takes place [25, 27], but if the N-H bond is not deactivated, an intramolecular cyclization occurs to give a new stabilized ylide [27, 28].

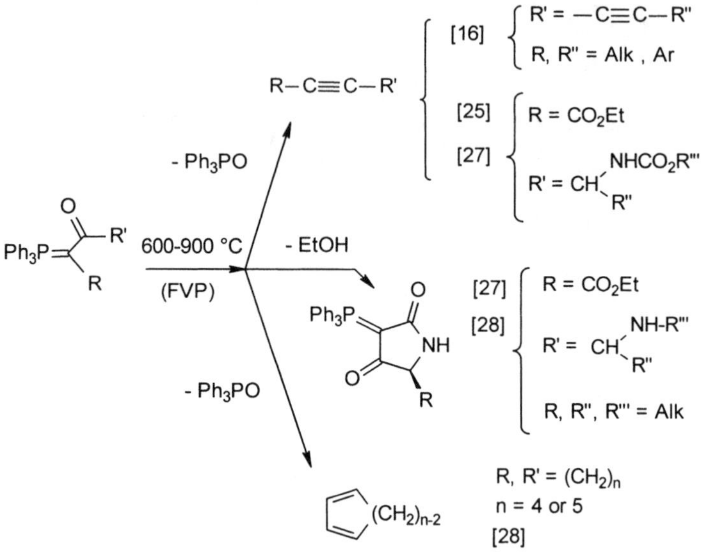

Scheme 5

It must also be pointed out that the alkyne synthesis through the FVP process can also be extended to bis(*β*-oxo phosphonium ylides) for the preparation of 1.3-diynes compounds [16].

2.2.2
Oxidation of Phosphonium Ylides

Oxidation is an important preparation process in the phosphonium ylide chemistry, which can be performed with a lot of oxidizing agents [29].

With *α-di*substituted ylides the reaction results in the formation of a carbonyl compound and a tertiary phosphine oxide (Scheme 6).

Interestingly, after oxidation of the *α*-cyano substituted ylides, the resulting *α*-ketonitriles can undergo a substitution of the cyano group, acting as a leaving group, by various amines to afford the corresponding amides. Wasserman and coworkers applied the sequence of reactions (acylation, oxidation, aminolysis) to the *α*-cyanomethyl phosphonium ylide in order to synthesize various peptidic biological active compounds [19, 20, 21, 24] or oxomalon diamides derivatives [23]. For their own part, Lee and Kawamura prepared 1,2,3-tricarbonyl compounds by oxidation of disubstituted stabilized ylides [30–32].

With *α-mono*substituted ylides the oxidation results in the formation of alkenes (by subsequent Wittig reaction on the intermediate aldehyde). A recent example of such synthesis is found in the preparation of all-(*Z*)-cyclododecatetraene by oxidation of the appropriate bis-ylide [33]. It must be pointed out that an approach of the same macrocycle based on ring closing metathesis was found ineffective.

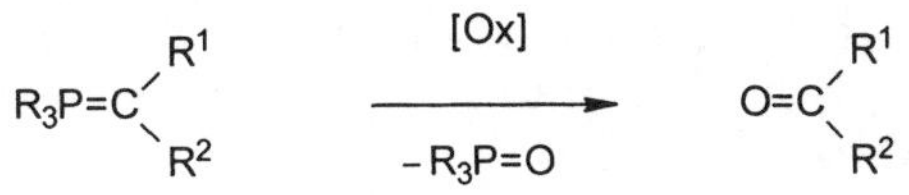

[Ox]	O_3 (CH_2Cl_2, - 78 °C)		O_2 (hν) (CH_3CN, 20 °C)	mCPBA ou MPP (CH_2Cl_2, - 25 °C)
R	Ph	Ph	*n*Bu	Ph
R^1	C(O)R'	C(O)R'	CO_2Me	C(O)R'
R^2	CN	C(O)NH-R'	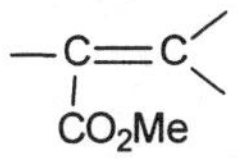	CO_2R
Ref	[19,20,21,23]	[22,24]	[30]	[31,32]

Scheme 6

2.2.3
Reaction of Stabilized Phosphonium Ylides with Activated C=C Double Bonds

D.K. Taylor and co-workers investigated thoroughly a new route to diastereomerically pure functionalized cyclopropanes utilizing stabilized phosphonium ylides and γ-hydroxyenones derived from 1,2-dioxines (Scheme 7) [34–38].

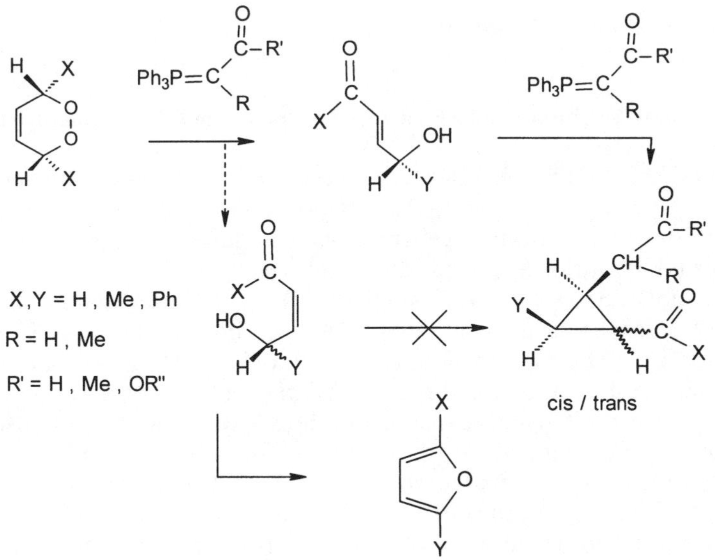

Scheme 7

Key features of the cyclopropanation include the ylide acting as a mild base to isomerize the 1,2-dioxines into *cis*-γ-hydroxy enones, followed by Michael addition of the ylide and last by cyclization of the intermediate enolate [35]. It must be noted that the *trans*-γ-hydroxyenones do not give the cyclopropanation.

In a related work, the same authors present an expeditious synthesis of functionalized dihydronaphtofurans starting from dihydronaphtodioxines and stabilized phosphonium ylides [39].

In the case of alkenes simply substituted by an electron-withdrawing group (without a γ-hydroxy group), the stabilized ylides give first a Michael addition and most often a subsequent prototropic shift resulting in new functionalized ylides (Scheme 8). Then a possible evolution of the resulting ylides can occur to give the final products [40–44].

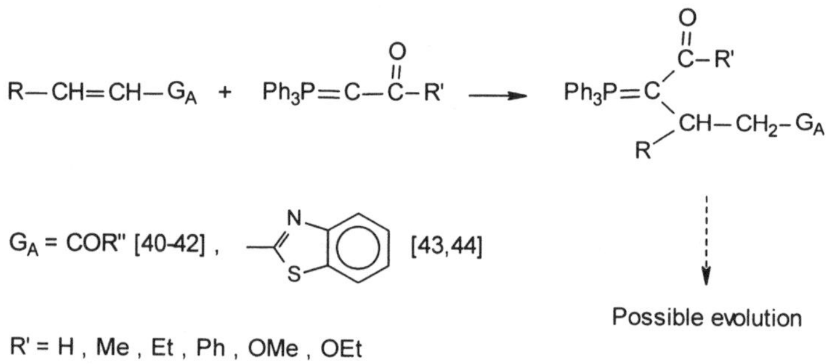

Scheme 8

3 C-Substituted Phosphonium Ylides

3.1 Ylides Substituted by Group 1 and 2 Elements; Phosphonium Yldiides and Diylides

The lithium phosphonium diylides **1**, first described by Wittig and Rieber (R=H) [45] were until recently mainly used as ligands in coordination chemistry [46]. These species also constitute excellent tools in organic synthesis [47], still recently attested by works described below (Scheme 9).

Thus unsubstituted (R=H) and substituted (R=alkyl) non-stabilized diylides **1** react with phenylisocyanate and dicyclohexylcarbodiimide (R^1NCX), leading to the formation of new monoylide type intermediates. These last ones react in situ with carbonyl compounds through a Wittig type reaction leading respectively to α,β-unsaturated amides **2** and amidines **3**, with a high *E* stereoselectivity, the double bond being di- or tri-substituted [48, 49]. By a similar reactional pathway, diylides also react with carbonic acid derivatives, with the synthesis as final products of *E*-α,β-unsaturated esters **4** and acids **5** [50].

Optically active vinyl sulfoxides **6** were also produced from the reaction of **1** (R=H) with(–)-(*S*)-menthyl sulfinate, via the intermediate formation with in-

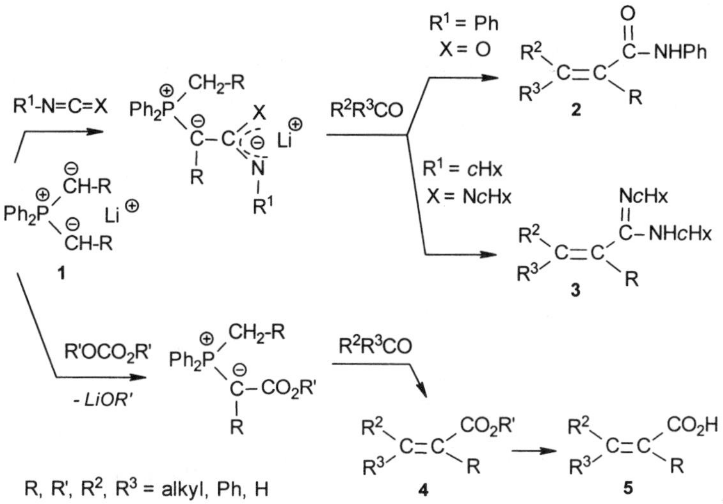

Scheme 9

version of configuration of the corresponding α-sulfinyl ylide and its subsequent reaction with carbonyl compounds (Scheme 10) [51].

The mixed lithium aminophosphonium azadiylides **7** [47] are known to be involved in reaction with various electrophiles specifically at the carbon center. The reaction with carbonyl known to form betaine adducts was recently used for the first time in Wittig type reactions for the synthesis of di-, tri-, and tetrasubstituted alkenes **8** [52].

Diylide **1**, by reaction with a phosphorus electrophile, Ph_2PCl, lead instantaneously via a nucleophilic substitution and intramolecular prototropy to the formation of functionalized monoylides **10** (Scheme 11).

In presence of carbonyl derivatives various Z-α,β-unsaturated phosphines **11** have been synthesized in very mild conditions [53–55]. The extension of this

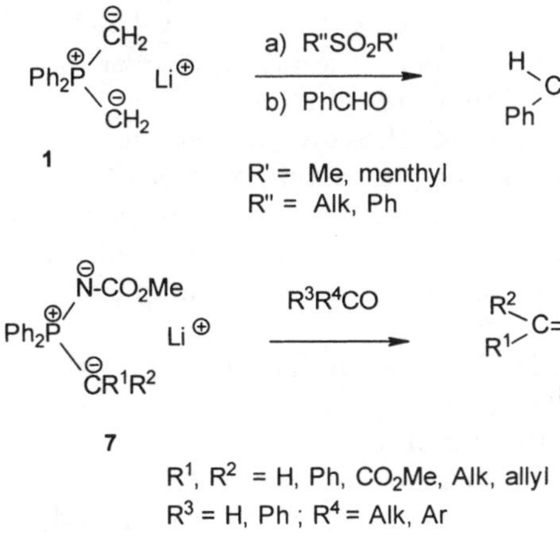

Scheme 10

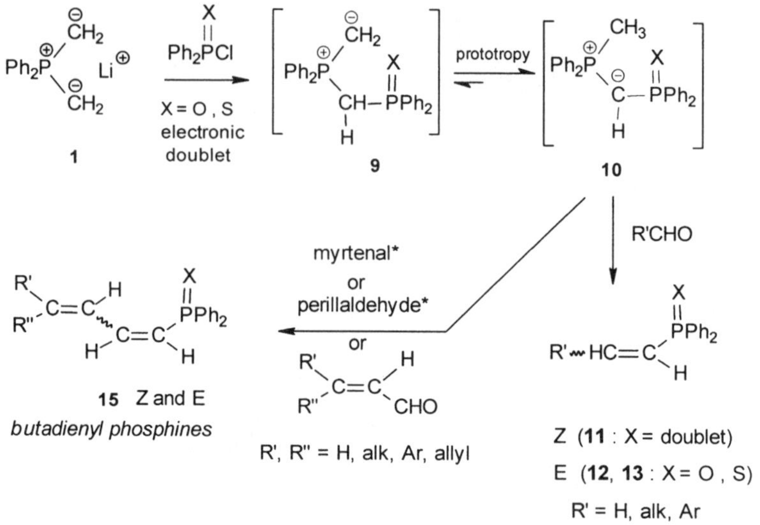

Scheme 11

study to other phosphorus electrophiles such as $Ph_2P(O)Cl$, $Ph_2P(S)Cl$, and $(EtO)_2P(O)Cl$ allowed the *E*-stereoselective synthesis of the corresponding styrylphosphine oxide **12**, sulfide **13** and dialkyl styrylphosphonates **14**. The method has been extended to a very efficient *one pot* synthesis of various butadienyl phosphines **15**, in some cases optically active, and difficult to obtain by other ways [56].

Recently the ylide **16** or the corresponding protonated ligands allowed, in presence of metallated bases of groups I and II (Li, K, Ba), the synthesis of the first phosphonium bridged metallocene **17** (dicyclopentadienylide) (Scheme 12). Chiral kalocene and barocene, observed only in racemic forms, have thus been obtained [57].

The barium phosphonium dibenzylide **18**, first organobarium compound not belonging to the cyclopentadienyl series, was isolated at the same time (Scheme 12) [58]. In the latter, the Ba^{2+} ion is encapsulated by two C_2-symmetric phosphonium ligands of opposite chirality.

At last, barium and calcium complexes of phosphonium bifluorenylide **19** was obtained from the corresponding phosphonium iodide (Scheme 12). The anion displays weak nucleophilicity toward both cations which prefer to coordinate with neutral oxygen of THF and with iodide. Located outside the coordination sphere of the metals, it represents thus the first example of uncomplexed phosphonium diylide [59].

As the phosphonium diylides, lithium phosphonium yldiides, first described by Schlosser and Corey ($Ph_3P{=}CR{-}Li$; R=H, C_3H_7) [60–62], have a high nucleophilicity and reactivity. Recently, the α-silylated lithium phosphonium yldiide **20** has been prepared from the stable phosphanyl-(silyl)carbene **19** and alkyllithium (Scheme 13). The first crystal X-ray diffraction study of such a reagent was proposed for **20** and its reaction with methyl iodide or phosphorus elec-

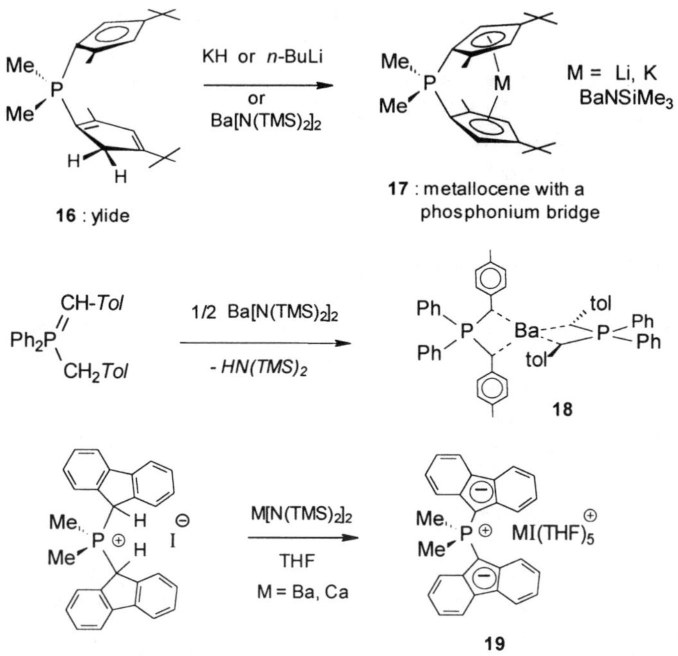

Scheme 12

trophiles allowed the synthesis of the corresponding C-substituted ylides [63]. Another yldiide **21**, of stabilized type, was also synthesized and its reaction with tin derivatives (R_3SnCl, R_2SnCl_2, or $SnCl_2$) resulted in the formation of two types of isomers corresponding to covalently *C*- and *O*-bonded tin substituted ylids (Scheme 13) (respectively **22** and **22′**) [64].

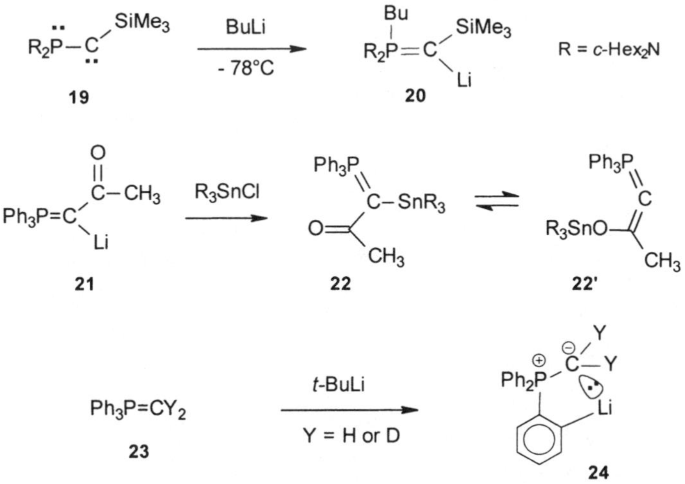

Scheme 13

Concerning the yldiide $Ph_3P{=}CH\text{-}Li$, the question as to whether it could be obtained by direct lithiation of the ylides **23** (Y=H), for a long time in debate between Schlosser and Corey, could have found an answer. Effectively NMR studies seem to show that a directed *ortho*-metallation occurs on an aromatic ring when **23** is reacted with *t*-BuLi [65], leading thus to the formation of **24** whose evolution does not afford $Ph_3P{=}CH\text{-}Li$ (Scheme 13).

3.2 Phosphonium Ylides in Coordination Chemistry

A review concerning the coordination and organometallic chemistry of P- and As-carbonyl stabilized ylides was published in1998. The references quoted inside, being previous to the period concerned by our article, are not detailed here but they represent a very interesting basis to the results described below [66].

3.2.1 *Ylides and Group 4 Transition Metals*

Complexation studies of phosphonium ylides and early transition metals of group 4 (titanium, zirconium, and hafnium), until now mainly concerning cyclopentadienyl species and non-stabilized ylides (Ph_3PCH_2), have been recently extended to α-keto-stabilized ylides **25** (Scheme 14). The latter (R=thiazolyl) in presence of $TiCl_4(THF)_2$ or $ZrCl_4$ forms new complexes **26** in which it acts as a ligand in a chelating manner through a N,O-coordination [67]. Compound **25** (R=Me, Ph) also reacts with tetrakis(dimethylamino) titanium or zirconium to give unstable complexes which evolve respectively to the formation of the new ylide **27** or the phosphinoxide **28**, via the transfer of a NMe_2 group [68]. From the reaction of the salt **29** with $Ti(NMe_2)_4$, another ylide was obtained in which the carbonyl oxygen binds the metal center [68].

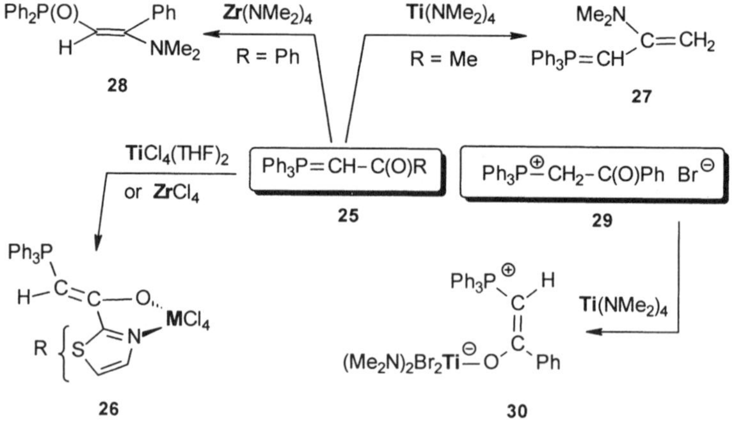

Scheme 14

Stable zirconate complexes of type **31** have also been described by the [3+2] cycloaddition of α-phosphino zirconaindene with alkyne derivatives, and the method has been extended to the synthesis of zirconium complexes from various heterocumulenes X=C=Y (X,Y=O, S, CyN) (Scheme 15) [69, 70].

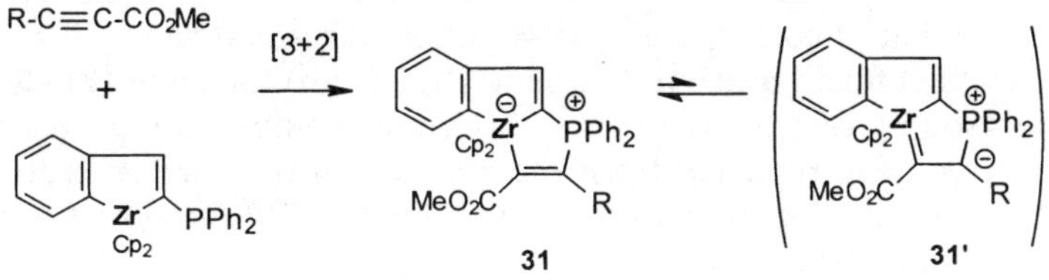

Scheme 15

3.2.2
Ylides and Group 5 Transition Metals

The first complexes of α-keto ylides and group 5 early transition metals have only recently been obtained by reaction of Nb(III) derivatives [{$NbCl_3$(dme)}(R′C≡CR″)] with **25** (R=thiazolyl) (Scheme 16). The chelation of the ylide occurs through an N,O-coordination to the metal center and in presence of MeLi a deprotonation of a phenyl ring takes place with the loss of alkyne, leading to the formation of a new *ortho*-metallated binuclear compound **32**. The two ylides involved in the complexation behave as tridentate anionic ligands and are mutually in a *trans* disposition in order to minimize the steric hindrance [71, 72]. Another binuclear niobium complex **33** has been obtained from **25** (R=Me, Ph) with this time an O-coordinated α-keto ylide [68].

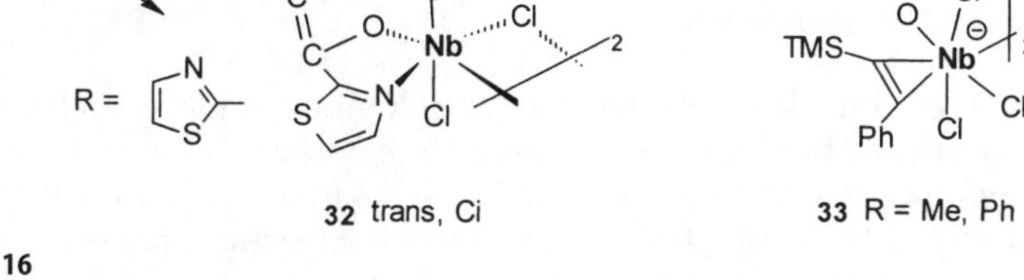

Scheme 16

3.2.3
Ylides and Group 6 and 7 Transition Metals

The diamagnetic ylide complexes **34** have been obtained from the reaction of electron-deficient complexes $[MoH(SR)_3(PMePh_2)]$ and alkynes (HC≡CTol for the scheme), via the formal insertion of the latter into the Mo–P bond. The structural data show that **34** corresponds to two different resonance-stabilized ylides forms **34a** (σ-vinyl form) and **34b** (carbene ylide form) (Scheme 17) [73]. Concerning the group 7 recent examples of *cis* ylide rhenium complexes **36** (*cis*-Me-Re-Me) have been reported from the reaction of the corresponding *trans* cationic alkyne derivatives **35** with PR''_3 *via* a nucleophilic attack of this phosphine at the alkyne carbon.

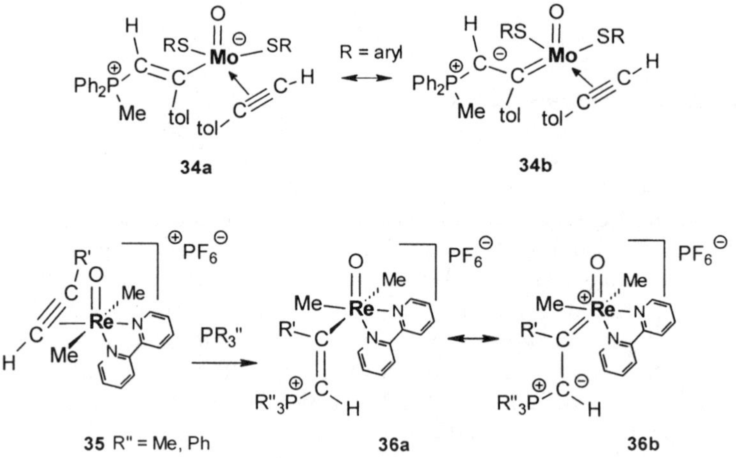

Scheme 17

As in the preceding case with molybdenum, the spectroscopic and X-ray crystallographic data suggest that the complexes obtained can be described as organometallic analogs of resonance-stabilized phosphonium ylides [74].

3.2.4
Ylides and Group 8 Transition Metals

New examples of dinuclear organoiron phosphonium ylide complexes, the species having already been known for about 20 years, have been obtained with the synthesis and X-ray structural characterization of iron compounds **37** (bridging ylide ligands) (Scheme 18) [75–77]. An interesting reaction corresponding to the first example of a cationic ligand migration to a cationic phosphenium phosphorus, has also been reported starting from another iron complex **38.** This one in presence of a Lewis acid (Me_3SiOTf) gives a dicationic phosphenium complex **39** and the intramolecular migration of the cationic phosphorus ylide occurs from the iron to the phosphenium (Scheme 18) [78].

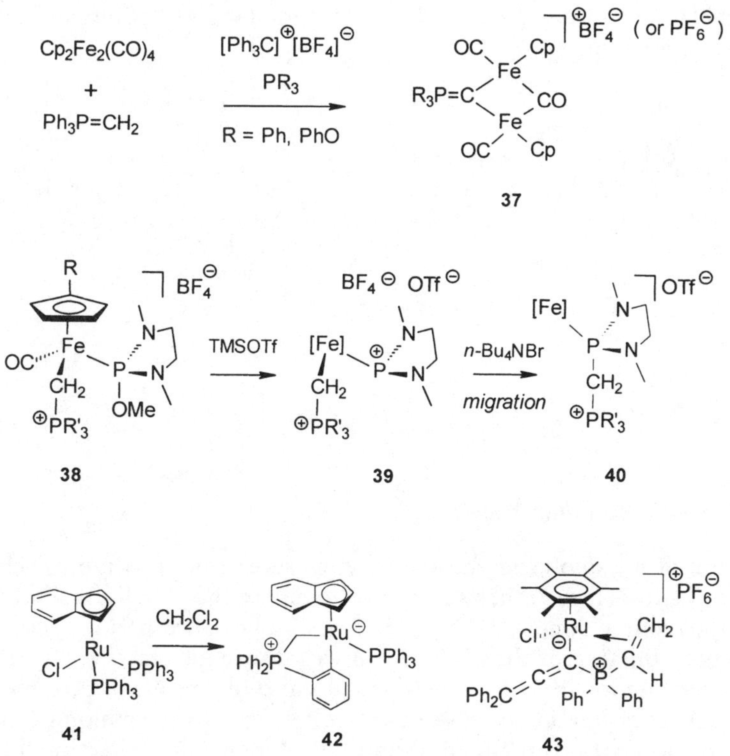

Scheme 18

However an unexpected new cyclic ruthenium phosphorus ylide half-sandwich complex **42** has been obtained by reaction of **41** with dichloromethane as solvent [79]. The cyclisation involves a C-Cl activation and corresponds to the incorporation of the methylene moiety in the P-C bond and to the ortho-metallation of one phenyl of the phosphine. An other novel unusual phosphonium ylide ruthenium complex **43** has also recently been described [80].

3.2.5
Ylides and Group 9 Transition Metals

A novel chiral dissymmetric chelating ligand, the non-stabilized phosphonium ylide of (R)-BINAP **44**, allowed in presence of $[Rh(cod)Cl]_2$ the synthesis of a new type of eight-membered metallacycle, the stable rhodium(I) complex **45**, interesting for its potential catalytic properties (Scheme 19) [81]. In contrast to the reactions of stabilized ylides with cyclooctadienyl palladium or platinum complexes (see Scheme 20), the cyclooctadiene is not attacked by the carbanionic center. Notice that the reactions of ester-stabilized phosphonium ylides of BINAP with rhodium(I) (and also with palladium(II)) complexes lead to the formation of the corresponding chelated compounds but this time with an equilibrium between a chelate form (described below with non-stabilized ylide) and an open and dicoordinated form [82].

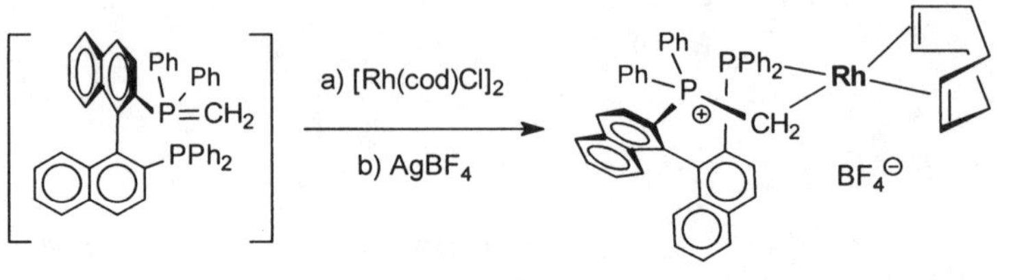

44 45

Scheme 19

3.2.6
Ylides and Group 10 Transition Metals

The reaction of the stabilized ylide **46** (α-vinyl substituted) with the cyclooctadienyl Pd(II) allows the synthesis of a novel complex, the (η^3-allyl)palladium **47**, in which the olefinic double bond participates in the coordination (Scheme 20) [83]. The coordination of the bis-ylide **48** with the same starting $Cl_2Pd(COD)$ leads to the formation of another new palladium complex **49** via COD exchange reactions. A C-coordination mode takes place between the carbanionic centers of the bis-ylide and the soft palladium and two stereogenic centers of the same configuration are thus created [83]. In contrast to the example described in Scheme 19, the $Cl_2M(COD)$ (M=Pd or Pt), in presence of a slightly different car-

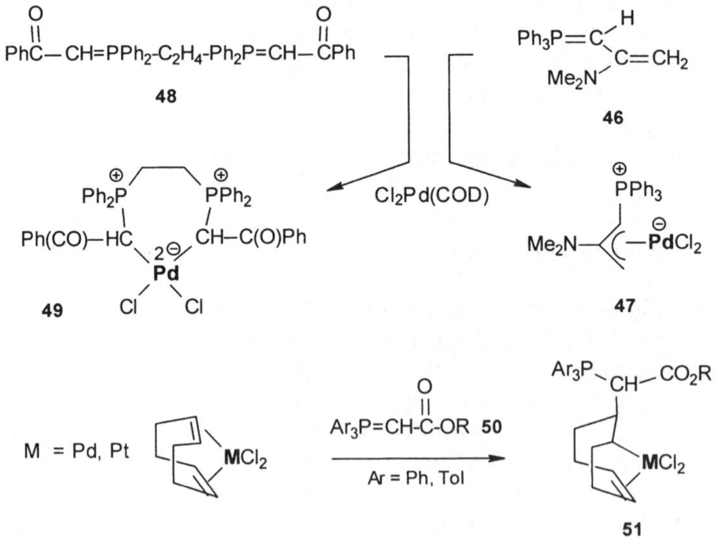

Scheme 20

bonyl-stabilized ylide **50**, gives a monomeric cyclooctenyl complex **51** resulting from the nucleophilic *exo*-attack of the COD by the ylide [84]. The formation of **51** constitutes the first example of a nucleophilic attack of an ylide at olefinic bond coordinated to Pd(II) or Pt(II).

Other works have shown that phosphonium ylides α-stabilized by a cyano or keto group can behave as ambidentate ligands towards palladium complexes (Scheme 21) [85–89].

Thus, the reaction of the cyclometallated palladium(II) **52** with the stabilized ylide Ph_3P=CH–C(O)Me gives the cationic complex **53** whose structure shows the presence of a O-bonded ylide *trans* to a soft arylic carbon of the dmba ligand. In contrast, the more nucleophilic ylide Ph_3P=CH–CO_2Me is selectively C-bonded (**54**) to the palladium via the ylidic carbon in *trans* position to the hard nitrogen of NMe_2. Finally, the cyano-stabilized ylide Ph_3P=CH-CN, depending on the stoichiometry of the reaction, leads to the mononuclear (or dinuclear) complex **55** in which both a C- and an N-coordination of the ylide are established (respectively *trans* to the NMe_2 and *trans* to the C_{Ar}).

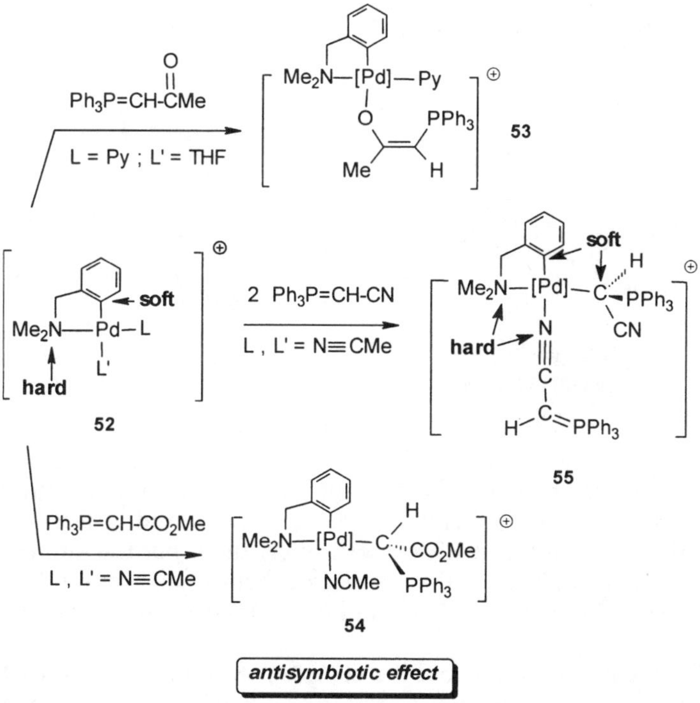

Scheme 21

The fact that the coordination of a hard "N" donor atom is preferred in a *trans* position to the soft "C" donor atom and reciprocally, constitutes a good illustration of the antisymbiotic effect which seems here to occur in presence of the soft Pd(II) metal. Similar observations have also been made in Pt(II) complexes.

The bis-phosphonium salt **56** in presence of $Pd(OAc)_2$ leads to the formation of the neutral bis-ylide **57** which reacts with $TlClO_4$ to give the dinuclear cationic complex **58** (Scheme 22) [89, 90]. The bis-ylide part, which has potentially two carbons and one oxygen donor atoms, acts as a C,C-chelating ligand through its two soft ylidic carbons.

Subsequent reactions involving **58** allowed the synthesis of various mono- or dinuclear complexes with ligands in *trans* position to the bis-ylide. All these ligands possess donor atoms behaving as borderline or hard bases (pyridines, N-bonded ylides, Cl^-, $acac^-$, NCMe, μ-OH). For example the reaction of the cyano-stabilized ylide $Ph_3P{=}CH{-}CN$ leads systematically to an N-coordination (**59**) and never to a C- one. Moreover strong C-donor ylides are unable to break the chlorine in **58** and the μ-hydroxo ligand in the dimer **62** is also strongly stabilized.

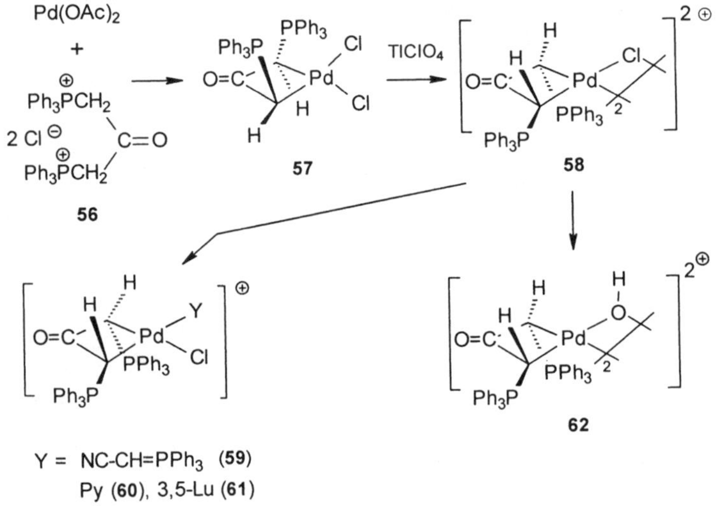

Scheme 22

Thus it appears that the presence of two soft carbons on the palladium stabilizes the *trans* coordination of hard ligands and drives the selective coordination of ambidente ligands through their hardest atom. These results, as those described in the previous scheme, constitute other examples of the antisymbiotic effect which can be observed in soft palladium(II) complexes.

It is interesting to notice that the complex **58** is also able to react with soft donor ligands such as triphenylphosphine resulting in the formation in very mild conditions of the unexpected orthometallated complex **59** (Scheme 23) [89, 91, 92]. The ligand here is linked to the metal through both an aromatic and an ylidic carbon. Other transformations are realized from **58** leading, including the compounds of the previous scheme, to four different structures for the bis-ylide: (i) C,C-chelate; (ii) C,C-orthometallated; (iii) C,C-orthometallated and free ylide; (iv) C,C,C-terdentate (Scheme 23).

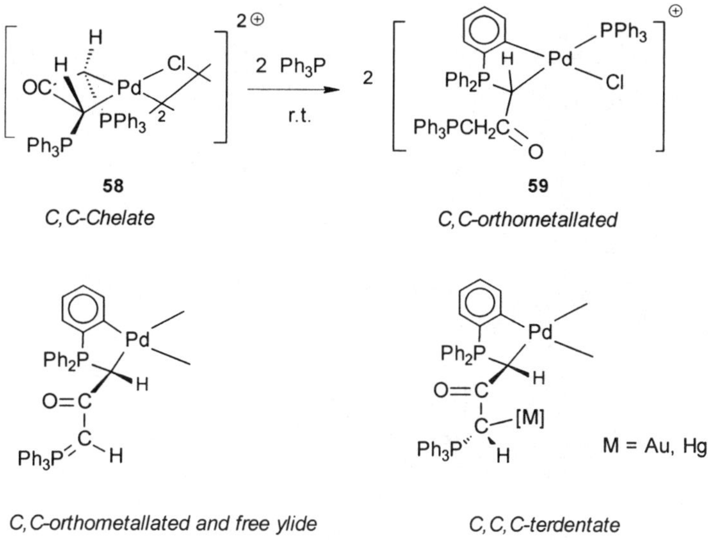

Scheme 23

Carbonyl-stabilized ylides and the cyano-stabilized ylides Ar_3P=CHCN in spite of their low nucleophilicity are able to give stable complexes. By comparison, very few complexes involving the second category of ylides are known. However a new example has recently been described by reaction of the cyano-ylide **60** with palladium(II) to give the complex *trans*-[$PdCl_2${CH($PTol_3$)CN}$_2$] **61** (Scheme 24) [93].

Others complexes (**63**), also recently described, constitute the first examples known of alkenyl phosphorus ylides of palladium (Scheme 24) [94]. The latter are obtained by reaction of the (dppm)Pd (II) **62** with various alkynes (substituted by electron-donating or electron-withdrawing groups). The reaction corresponds to an unprecedented apparent insertion of alkynes into a palladium-phosphine bond. This insertion is highly regiospecific since only alkenyl ylides with the terminal alkyne carbon bonded to the phosphorus center are formed. The formation of this thermodynamically stable alkenyl ylide is potentially interesting in the field of deactivation catalysis for palladium catalyzed organic transformations using unsaturated hydrocarbons.

A related unprecedented double insertion of electron-deficient alkynes has also been reported in the reactions of the linear Pt_2Pd heterotrimetallic complex **64** with **65** ($RO_2CC{\equiv}CR$) (Scheme 24) [95, 96]. A series of unsymmetrical A-frame clusters **68** has thus been obtained in which a first insertion of the alkyne takes place site-selectively into the Pt-Pd bond vs the Pt-Pt bond (**66**). After a zwitter-ionic polar activation of the resulting inserted alkene (**67**), a subsequent reaction with the phosphine unit of the dpmp allows one to obtain the products **68** via the nucleophilic migration of the terminal P atom from the Pd center to the CH terminal carbon (formation of the P-C bond).

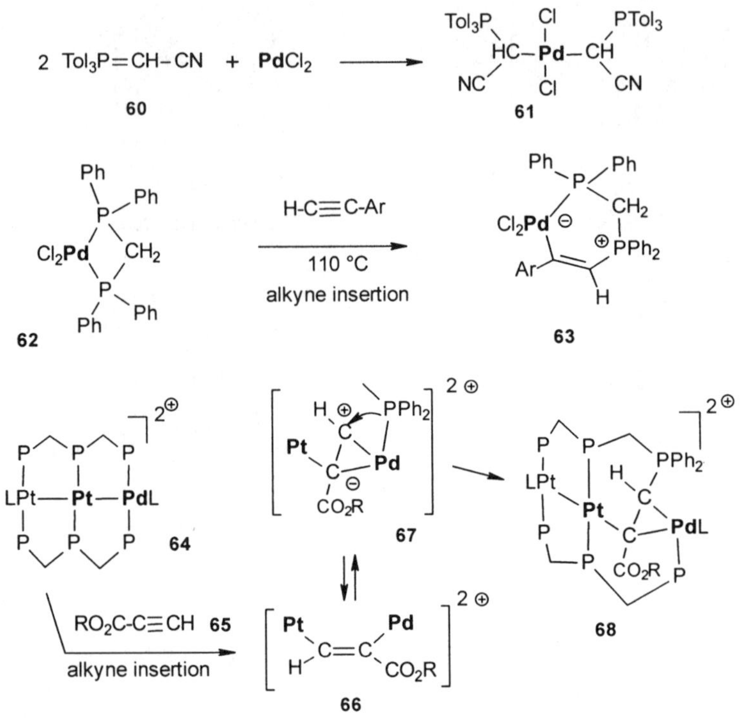

Scheme 24

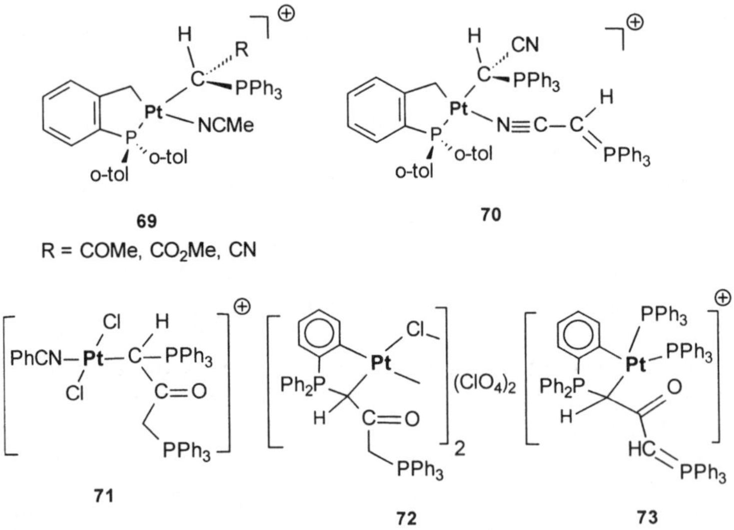

Scheme 25

Works developed by Navarro and coworkers for the palladium and stabilized ylides complexes have been extended to the platinum chemistry (Scheme 25) [97, 98].

Related platinum compounds such as **69** have thus been synthesized from α-keto or cyano-stabilized ylides. A selective *C*-coordination *trans* to the P phosphorus atom of the ligand is observed, probably resulting from steric interactions.

The compound **70** has also been reported showing the ambident character (both *C*- and *N*-coordination) of the cyano-stabilized ylide as ligand. The authors have also transposed their work concerning the keto-bis-ylide and palladium, with the synthesis of the C-bonded complex **71** or the new cycloplatinated orthometallated compound **72**. The latter by various treatments allows one to obtain other ylidic cationic complexes of platinum such as **73**. A *C,C,C*-terdentate coordination of the keto bis-ylide, already observed with the palladium is also obtained from the reaction of **73** with gold derivatives.

3.2.7
Ylides and Group 11 Transition Metals

Complexes of phosphorus ylides with gold have been known for a long time [29, 46]. Two related reviews, recently published, are especially dedicated to dinuclear gold species held together in most of the cases by the diphenylphosphonium diylide ligand, the complex **74** being by far the most studied (Scheme 26) [99, 100]. A very rich chemistry has been developed around these diauracycle systems and the objective of one of these reviews concerns the implications of these species for the understanding of oxidative addition processes in dinuclear and polynuclear complexes [100]. The second review is more particularly dedicated to the reactivity of the diauracycle systems and their implication in the preparation of chains of gold atoms [99].

Recent works have been reported in the field of these dinuclear complexes. Thus new products of type **76** have been described resulting from the substitution of brominated dinuclear gold(II) and gold(III) diylides complexes by the lithium methylenediphenylthiophosphinate $Li[CH_2P(S)Ph_2]$ (Scheme 26). For example starting from **75** (R=Me) a one-electron reduction of each gold atom of the Au^{III}/Au^{III} metallacycle occurs to produce the corresponding homovalent Au^{II}/Au^{II} metal-metal bonded complexes **76** [101].

A related dinuclear species **77**, recently described, constitutes the first dinuclear gold(I) complex with heterobridged phosphor-1,1-dithiolato moieties and bis(ylide) bridging ligands [102]. It is obtained by reaction between $[AuS_2PPh_2]$ and the diylide gold complex **74** (R=Me). No intermolecular Au-Au interaction is observed in **77** but the oxidative addition of chlorine to the product leads to a new complex **78** in which a single bond is formed between the two Au(II) centers (Scheme 26).

Very few examples of bridging non-cyclic methanides of gold are known. Among them the complex **79** has been reported as the result of the reaction of phosphine-phosphonium derivatives with acetylacetonate derivatives of gold(I) [103, 104]. The complexes **80** [89, 98], already seen in previous paragraphs, cor-

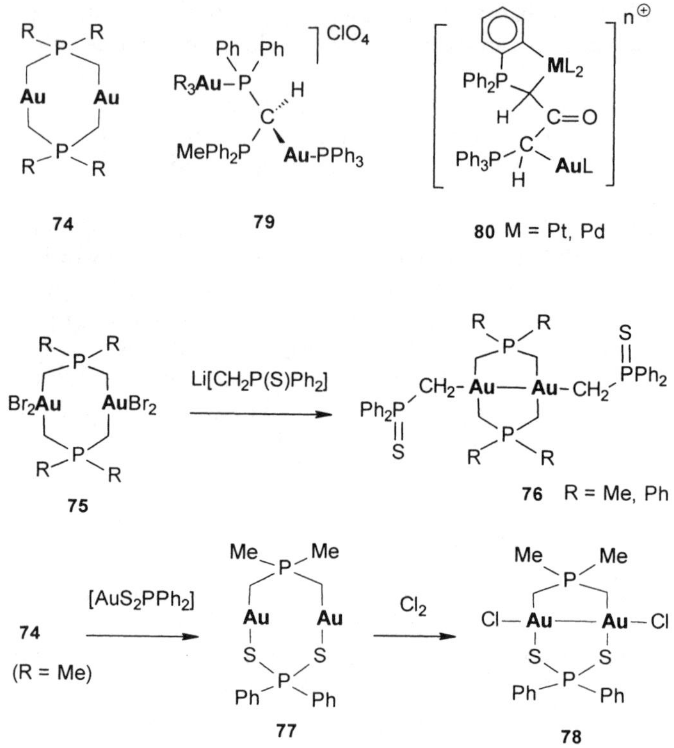

Scheme 26

respond to a *C,C,C*-terdentate coordination of a keto bis-ylide both to a gold atom and to a palladium or a platinum one. Finally another compound has been described which also combines the coordination to two different metals. It consists of a gold eight-membered diauracycle linked to two tripodal *tris*(amido)tin fragments [105].

3.2.8
Ylides and Group 12 Elements

When the phosphonium ylide **81** is reacted with zinc amide, the corresponding α-zincated phosphorus ylide is formed. Thermally unstable, it evolves almost quantitatively to zincatacyclobutane **82** which in presence of pyridine leads to the formation of the zincataphosphoniaindane **83**. In order to explain this unprecedented cyclometallation reaction, a mechanism is proposed involving a low coordinated zinc center. The new product, reacted with benzaldehyde leads to the diphenylallene **84** (Scheme 27) [106–108].

Usually, *C*-mercury substituted phosphorus ylides are monomers and in order to stabilize these complexes the presence of a second substituent on the carbon is necessary to balance the electron-donating effect of the metal. However a dimeric complex **85** has been obtained by the reaction of mercuric halides HgX_2

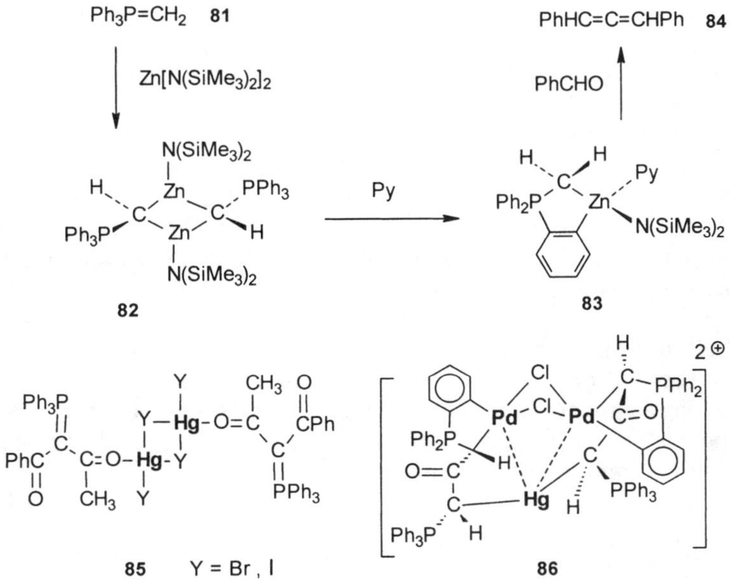

Scheme 27

with a diketo-ylide (Scheme 27) [109]. The latter is regiospecifically coordinated to the soft mercuric center via the acetyl oxygen and this is the first example of an O-coordination of any keto phosphorus ylide to Hg(II).

The compound **86** constitutes an interesting compound in which a bis keto-ylide coordinates both a palladium (through an aryl and an ylidic carbon) and a mercuric center (through an ylidic carbon) [89, 110]. This *C,C,C*-terdentate coordination has also been observed with gold complexes.

3.3
Phosphonium Ylides Substituted by Elements of Groups 13–16

3.3.1
Ylides C-Substituted by Elements of Group 13–14

Tris(pentafluorophenyl)borane $B(C_6F_5)_3$, like the other compounds of group 13, is a strong Lewis acid which can react with an ylide such as **87** to give the corresponding adduct **88**. The formation of the latter is reversible and under prolonged thermolysis it evolves to the thermodynamic product **89** which results from the ylide nucleophilic attack at one of the $-C_6F_5$ rings followed by a fluorine transfer (Scheme 28) [111].

For group 14, *C*-tin-substituted phosphorus ylides have been studied in the past but less than the corresponding *C*-silyl ylides which, owing to their stability and reactivity, are of considerable interest. Indeed a recent review concerning the silylphosphanes with one part concerning the synthesis and applications of silylated phosphorus ylides has been published [112].

As a recent result an example of *C*- and *O*-covalently bonded tin substituted ylides (respectively **90** and **91**) has been reported, the adducts resulting from the reaction of a stabilized yldiide with tin derivatives (R_3SnCl, R_2SnCl_2, or $SnCl_2$) (Scheme 28) [64].

Some ylides **92** among them *C*-silylated ones have been synthesized in order to compare the stabilization influence on the carbanionic center of various *C*-substituents (I, $SiMe_3$, Ph_3P^+) [113]. It appears from this study that the stabilization due to the electron-withdrawing *C*-substituents (R or R′) is not so negligible by comparison with the hyperconjugative stabilization between the ylidic carbon and the phosphonium group [114]. This is particularly true for the iodine substituent.

Triphenylphosphonium ylide reacts with the silylene complex **93** which has a highly electrophilic silicon center, to give the corresponding cationic adduct **94** [115]. The lengthening of the PC bond indicates a loss of the double bond character of the ylide and corresponds to the formation of a tetrahedral silicon center with four covalent bonds (Scheme 28).

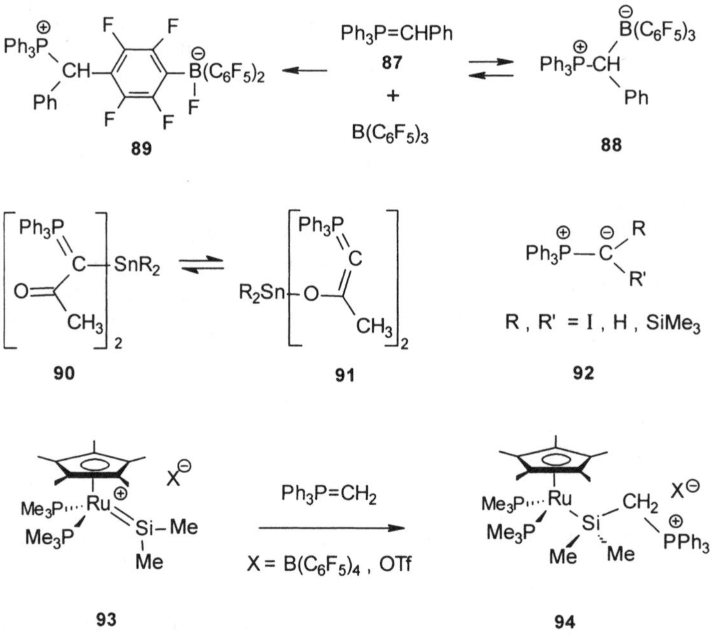

Scheme 28

Trimethylsilyl ylide **95** reacts with PCl_3 or $AsCl_3$ to give respectively the corresponding halophosphorus ylide **96** and the first example known of haloarsanyl ylide **97** [116]. In various stoichiometric conditions the 1:1 condensation of **95** with $AsCl_3$ is realized to give haloarsanyl ylides with either a dimeric structure (diarsenate **98**) or trimeric and tetrameric cationic compounds **99** and **100** (Scheme 29).

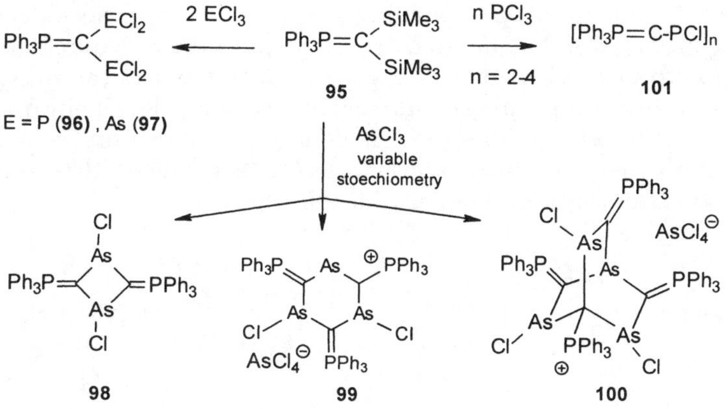

Scheme 29

A corresponding chemistry with the formation of oligomeric halophosphanyl ylides **101** has been developed by the same authors [117, 118].

The symmetrical bis(ylidyl)phosphenium chlorides **103**, obtained from the reaction of trimethylsilyl ylides **102** with PCl_3 are the first phosphenium salts which do not need counterions of low basicity such as $AlCl_4^-$ to be isolated (Scheme 30) [119]. The explanation of their stability lies in the delocalisation of the phosphenium charge in the two phosphonium parts. The reactivity study of these species is reported and for example the phosphenium **103** (R=Ph) adds ortho quinones to the central phosphorus to give the corresponding dioxaphospholenium salts **104** via a [4+1] cycloaddition.

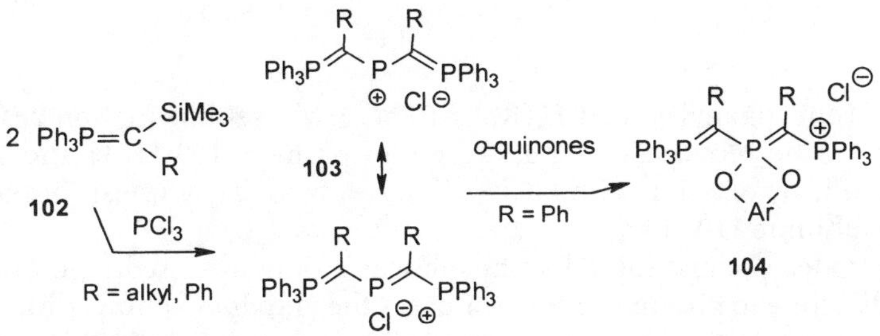

Scheme 30

3.3.2
Ylides C-Substituted by Elements of Group 15

Important literature is available for this type of ylides which are usually thermally stable in the case of phosphorus, arsenic or stilbene *C*-substituents. This is different for ylides *C*-substituted by nitrogen atoms which have a destabilizing effect.

As recent results we can mention the reactions of ylides **105** in presence of PCl_3, $AsCl_3$, or $SbCl_3$ which lead to the formation of 1:1 adducts (Scheme 31)

[120]. With PCl_3 the form corresponding to a dissociation of the adduct in two ions is obtained (**106′**). Starting from $AsCl_3$ structures **106** with different degrees of transition from a zwitterionic *ψ*-trigonal-bipyramide to a cationic *ψ*-tetrahedron are observed depending on the substituents R and R′. Finally from $SbCl_3$ an asymmetric dimer type **106** whose structure is close to a square-pyramidal geometry is obtained. Dicationic phosphorus (**107**) or arsenic (**108**) adducts are also described from various types of ylides **105** (Scheme 31).

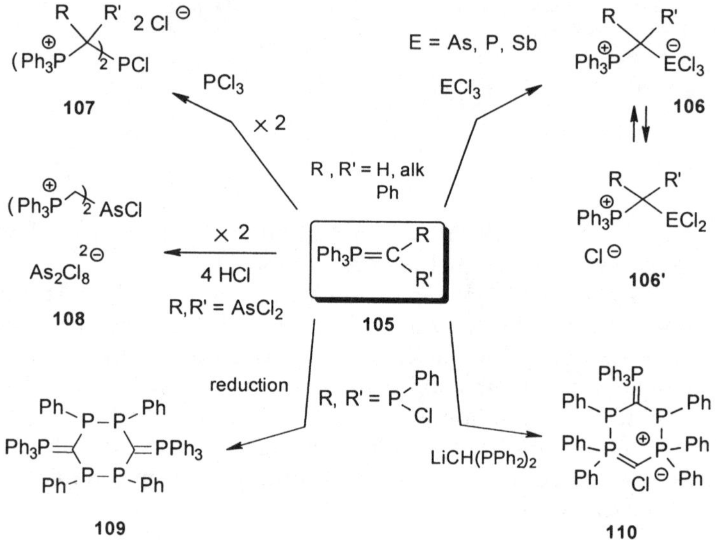

Scheme 31

The bis(phosphanyl)ylide **105** (R,R′=PPhCl), whose chlorination destroys the chirality can be reduced in 1,2,4,5-tetraphosphinane **109** [121] and the same starting ylide by reaction with $LiCH(PPh_2)_2$ leads to the synthesis of the 1,2,4,5-tetraphosphinine **110** [122].

Other ylides *C*-substituted by phosphorus atoms, after reduction react with carbon disulfide to give in presence of bases the ylidylphosphinyl dithioformate **112** (Scheme 32) [123]. The same starting ylide in presence of $PtCl_2$ gives the corresponding platinum complex (two diastereomers) [123] and with trimethylsilyl phosphines, ylides **114** substituted by diphosphines are obtained. A cyclisation product is also observed when **111** (R=CH_2Cl) is reacted with 2 equiv. $LiN(PPh_2)_2$ to give the 1,2,3,5-azatriphosphole **115** (Scheme 32) [122]. Other works concerning the study of the reactivity of triphenylphosphonium allylides toward PCl_3 have been reported, thus allowing the synthesis of new examples of 1,2-diphospholes and of ionic phenalene structures [124].

Various compounds of type **116** have been obtained by the reaction of PF_4 with bidentate ligands incorporating a bridging >C=PPh_3 group (Scheme 33) [125]. Thanks to the increased electron density of the latter the chelating ability of these ligands is better than the corresponding bases with >C=CH_2 and >CHX bridges.

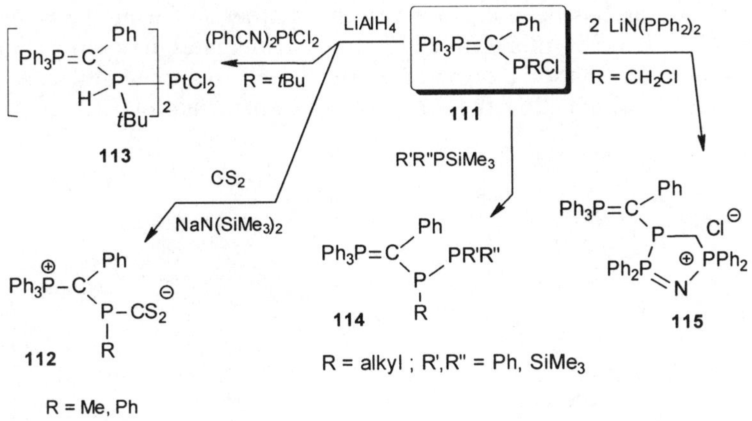

Scheme 32

Biphilic ylides 117 can enter in cyclocondensations with carboxylic acid chlorides, carbon disulfide, and acyl isothiocyanates [126]. Certain corresponding heterocyclic products **118** obtained are described for the first time (Y′=N(CO)Ph; Y=S) and are precursors of the new 1-(2-phenylthiazol-5-yl)-5-phenyltetrazole **119** (Scheme 33).

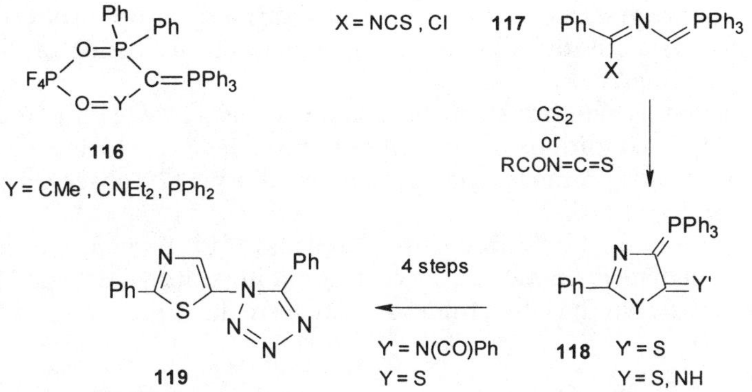

Scheme 33

3.3.3
Ylides C-Substituted by Elements of Group 16

Phosphorus ylides *C*-substituted and stabilized by elements of group 16 are often used for the synthesis of natural substances. For example, the synthesis of simplified analogs of artemisinin, used against chloroquine-resistant malaria, has been recently described from methoxymethylphosphonium ylide **120** [127, 128]. The later is able to convert aliphatic nitriles into α-functionalized ketones **122** which are the precursors of the target compounds. Starting from the aromatic nitriles the reaction has been extended to the synthesis of α-methoxyacetophenones **124** which are of interest as synthetic intermediates in organic chemistry. This method offers a good alternative to the use of organometallic compounds (tin derivatives) which allow the same type of transformation (Scheme 34).

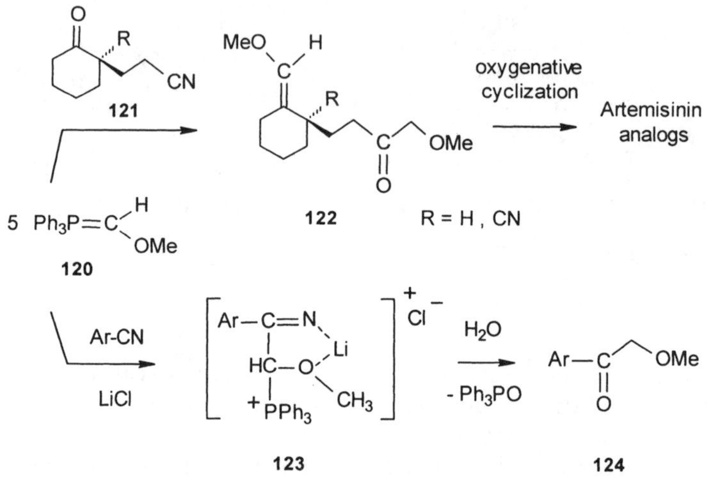

Scheme 34

In Sect. 2.2.1 we described the thermolysis of various β-oxophosphorus ylides affording either substituted alkynes or cyclic dienes by extrusion of Ph_3PO, or new stabilized ylides.

In contrast to this behavior stabilized sulfonyl ylides **125** lose Ph_3P under flash vacuum pyrolysis conditions to give sulfonyl carbenes **127** which evolve to the formation of various products among which the alkenes **129**, observed in all the cases (Scheme 35) [129].

Starting from sulfinyl stabilized ylides **126**, a loss of Ph_3P was also observed to give the intermediate formation of the corresponding sulfinyl carbene **128**. Via a 1,2-oxygen transfer in the latter, thioesters **130** were this time obtained [130].

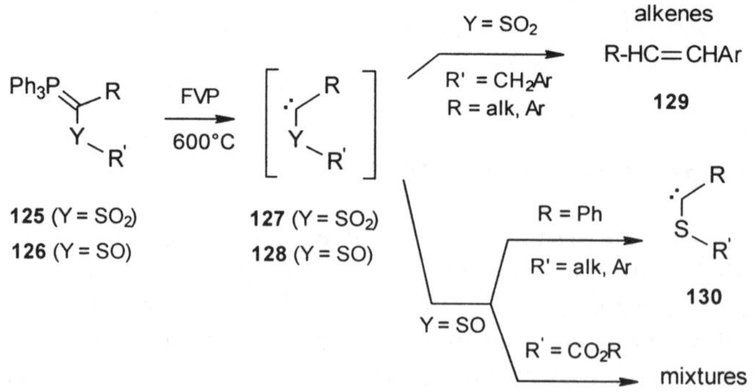

Scheme 35

Finally the new ylides stabilized with sulfinyl but alkoxy carbonyl groups also behave in a more complex manner than for the simple sulfinyl ylide and the loss of both Ph_3PO and Ph_3P is observed [131]. Mixtures of products are obtained among which are alkenyl sulfides as major products but also sulfides and thioesters (Scheme 35).

The cycloaddition of alkynes with the tributylphosphine-carbondisulfide adduct **131** results in the in situ formation of the ylides **132** which react with aldehydes to give the novel 2-arylidene or 2-alkylidene-1,3-dithioles **133** (Scheme 36) [132]. Concerning ylides *C*-substituted by sulfur we can also mention a publication on the behavior of various keto-stabilized ylides towards acyclic and cyclic *cis*-disulfides allowing the synthesis of substituted thiazoles, thiols, and dithiols [133].

Finely symmetrical divinyl tellurides **135**, especially useful in transmetallation reactions, have been prepared in situ by sequential reaction of an excess of the ylide **134** with $TeCl_4$ and aldehydes (Scheme 36) [134]. These compounds whatever their geometry lead in presence of *n*-BuLi to the formation of *E*-α,β-unsaturated aldehydes.

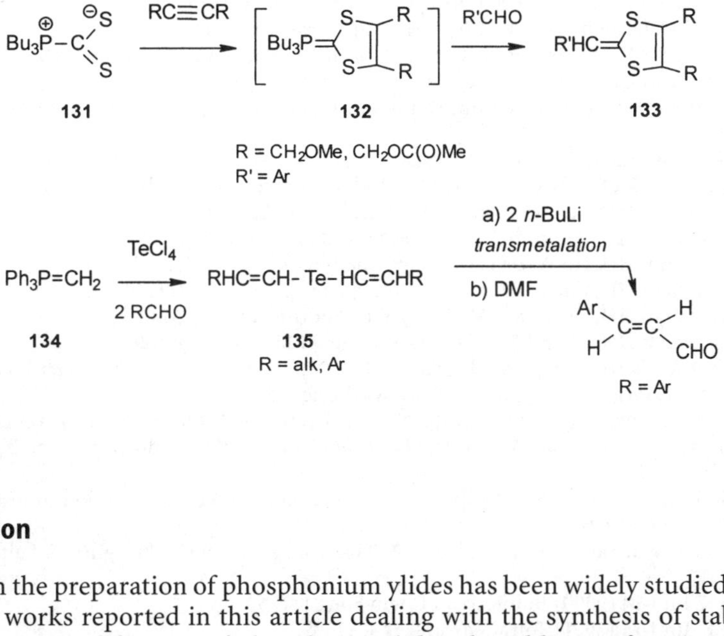

Scheme 36

4
Conclusion

Although the preparation of phosphonium ylides has been widely studied in the past, the works reported in this article dealing with the synthesis of stabilized ylides by the modification of phosphorus ylides, the addition of tertiary phosphines to multiple bonds or the modification of the side chain, still attest to the current importance of this subject. Moreover the study of the reactivity of these stabilized ylides with their thermolysis, their oxidation (synthesis of ketones), or their reaction with activated double bonds (cyclopropanation reaction) still enlarges the application field of these reagents in organic synthesis. Concerning phosphorus ylides *C*-substituted by heteroatoms of groups 1–16, the recent works presented in this article show for this reagent a very large application domain in organic synthesis as well as in coordination chemistry. Taking into account the extreme variability of *C*-substitution, a very promising investigation area can still be expected for this type of ylide.

5
References

1. Ramazani A, Shajari N, Gouranlou F (2001) Phosphorus Sulfur Silicon 174:223
2. Yavari I, Adib M (2001) Tetrahedron 57:5873
3. Yavari I, Maghsoodlou MT, Djahaniani H, Hazeri N (1999) J Chem Res (S) 216
4. Baharfar R, Heydari A, Saffarian N (2001) J Chem Res (S) 536
5. Shaabani A, Safaei HR, Bijanzadeh HR (2001) Synth Commun 31:2639
6. Yavari I, Zabarjad-Shiraz N (2001) Phosphorus Sulfur Silicon 176:141
7. Yavari I, Nourmohammadian F (2000) Tetrahedron 56:5221
8. Ramazani A, Mohammadi-Vala M (2001) Phosphorus Sulfur Silicon 176:243
9. Chuang S-C, Santhosh KC, Lin C-H, Wang S-L, Cheng C-H (1999) J Org Chem 64:6664
10. Hsia T-Y, Chidambareswaran SK, Cheng C-H (1998) J Org Chem 63:6119
11. Keglevich G, Forintos H, Körtvélyesi T, Töke L (2002) J Chem Soc Perkin Trans 1 26
12. Keglevich G, Körtvélyesi T, Forintos H, Vaskó AG, Vladiszlav I, Töke L (2002) Tetrahedron 58:3721
13. Boulos LS, Arsanious MHN (1999) Heteroatom Chem 10:337
14. Pascoly G, Lavendy H (2001) Int J Mass Spectrom 206:153
15. Zhdankin VV, Maydanovych O, Herschbach BJ, Matveeva ED, Zefirov NS (2002) Tetrahedron Lett 43:2359
16. Aitken RA, Hérion H, Horsburgh CER, Karodia N, Seth S (1996) J Chem Soc Perkin Trans 1 485
17. Aitken RA, Burns G, Moison JJ (1998) J Chem Soc Perkin Trans 1 3937
18. Meshram HM, Reddy GS, Reddy MM, Yadav JS (1998) Tetrahedron Lett 39:4107
19. Wasserman HH, Petersen AK (1997) Tetrahedron Lett 38:953
20. Wasserman HH, Petersen AK (1997) J Org Chem 62:8972
21. Wasserman HH, Ho W-B (1994) J Org Chem 59:4364
22. Wasserman HH, Chen J-H, Xia M (1999) J Am Chem Soc 121:1401
23. Wasserman HH, Lee K, Xia M (2000) Tetrahedron Lett 41:2511
24. Wasserman HH, Chen J-H, Xia M (2000) Helv Chim Acta 83:2607
25. Aitken RA, Karodia N, Massil T, Young RJ (2002) J Chem Soc Perkin Trans 1 533
26. Eymery F, Iorga B, Savignac P (2000) Synthesis 185
27. Aitken RA, Buchanan GM, Karodia N, Massil T, Young RJ (2001) Tetrahedron Lett 42:141
28. Aitken RA, Bjørnstad V, Massil T, Skramstad J (1999) Phosphorus Sulfur Silicon 144/146:577
29. Kolodiazhnyi OI (1999) Phosphorus ylides: chemistry and application in organic synthesis. Wiley-VCH, p 73
30. Kawamura Y, Sato Y, Arai M, Horie T, Tsukayama M (1999) Phosphorus Sulfur Silicon 149:1
31. Lee K, Im J-M (2000) Bull Korean Chem Soc 21:1263
32. Lee K, Im J-M (2001) Tetrahedron Lett 42:1539
33. Guiard S, Santelli M, Parrain J-L (2001) Synlett 553
34. Avery TD, Haselgrove TD, Rathbone TJ, Taylor DK, Tiekink ERT (1998) Chem Commun 333
35. Avery TD, Taylor DK, Tiekink ERT (2000) J Org Chem 65:5531
36. Avery TD, Greatrex BW, Taylor DK, Tiekink ERT (2000) J Chem Soc Perkin Trans 1 1319
37. Palmer FN, Taylor DK (2000) J Chem Soc Perkin Trans 1 1323
38. Avery TD, Fallon G, Greatrex BW, Pyke SM, Taylor DK, Tiekink ERT (2001) J Org Chem 66:7955
39. Haselgrove TD, Jevric M, Taylor DK, Tiekink ERT (1999) Tetrahedron 55:14,739

40. Abdou WM, Salem MI, Sediek AA (1998) J Chem Research (S) 28
41. Boulos LS, Shabana R, Shaker YM (2000) Heteroatom Chem 11:57
42. Moorhoff CM (2000) Recent Res Devel Organic Chem 4:239
43. Abdou WM, Ganoub NA, Shaddy AAM (1998) Tetrahedron 54:9079
44. Ganoub NA (1999) Phosphorus Sulfur Silicon 148:21
45. Wittig G, Rieber M (1949) Justus Liebigs Ann Chem 562:177
46. Schmidbaur H (1983) Angew Chem Int Ed Engl 22:907
47. Cristau HJ (1994) Chem Rev 94:1299
48. Cristau HJ, Taillefer M, Urbani JP, Fruchier A (1996) Tetrahedron 52:2005
49. Cristau HJ, Taillefer M, Urbani JP (1996) Phosphorus Sulfur Silicon 111:123
50. Cristau HJ, Taillefer M (1998) Tetrahedron 54:1507
51. Mikolajczyk M, Perlikowska W, Omelanczuk J, Cristau HJ, Perraud-Darcy A (1998) J Org Chem 63:9716
52. Peralta Pérez E, Lopez Ortiz F (2000) Chem Commun 2029
53. Taillefer M, Cristau HJ (1998) Tetrahedron Lett 39:7857
54. Taillefer M, Cristau HJ, Jouanin I, Inguimbert N, Jäger L (1999) Phosphorus Sulfur Silicon 144:401
55. Taillefer M, Cristau HJ, Fruchier A, Vicente V (2001) J Organomet Chem 624:307
56. Taillefer M, Cristau HJ, Fruchier A, Vicente V (2003) Eur J Org Chem (to be published)
57. Leyser N, Schmidt K, Brintzinger HH (1998) Organometallics 17:2155
58. Harder S, Lutz M (1997) Organometallics 16:225
59. Brady ED, Hanusa TP, Pink M, Young VG (2000) Inorg Chem 39:6028
60. Schlosser M, Steinhoff G, Kadibelban T (1966) Angew Chem Int Ed Engl 5:968
61. Corey EJ, Kang J (1982) J Am Chem Soc 104:4724
62. Schlosser M, BaTuong H, Respondek J, Schaub B (1983) Chimia 37:10
63. Goumri-Magnet S, Gornitzka H, Baceiredo A, Bertrand G (1999) Angew Chem Int Ed Engl 38:678
64. Khalil MI (1998) Main Group Met Chem 21:193
65. Korth K, Sundermeyer J (2000) Tetrahedron Lett 41:5461
66. Belluco U, Michelin RA, Mozzon M, Bertani R, Facchin G, Zanotto L, Pandolfo L (1998) J Organomet Chem 557:37
67. Otero A, Fernandez-Baeza J, Antinolo A, Carrillo-Hermosilla F, Tejeda J, Lara-Sanchez A, Lopez-Solera I (2001) J Organomet Chem 629:68
68. Spannenberg A, Baumann W, Rosenthal U (2000) Appl Organometal Chem 14:611
69. Miquel Y, Igau A, Donnadieu B, Majoral J-P, Pirio N, Meunier P (1998) J Am Chem Soc 120:3504
70. Cadierno V, Zablocka M, Donnadieu B, Igau A, Majoral J-P (1999) Organometallics 18:1882
71. Diez-Barra E, Fernandez-Baeza J, Kubicki MM, Lara-Sanchez A, Otero A, Poujaud N, Tejeda J (1997) J Organomet Chem 542:291
72. Antinolo A, Carrillo-Hermosilla F, Diez-Barra E, Fernandez-Baeza J, Lara-Sanchez A, Otero A, Tejeda J (1998) J Organomet Chem 570:97
73. Fairhurst SA, Hughes DL, Marjani K, Richards RL (1998) J Chem Soc Dalton Trans 1 899
74. Jung J-H, Hoffman DM, Lee TR (2000) J Organomet Chem 599:112
75. Churchill MR, Korzenski MB, Janik TS (1996) J Chem Cryst 26:683
76. Churchill DG, Churchill MR, Janik TS (1997) J Chem Cryst 27:667
77. Churchill MR, Keil KM, Janik TS, Willoughby MC (2001) J Coord Chem 52:229
78. Nakazawa H, Kishishita M, Nakamoto T, Nakamura N, Ishiyama T, Miyoshi K (2000) Chem Lett 230
79. Park JH, Koh JH, Park J (2001) Organometallics 20:1892
80. Hansen HD, Nelson JH (2000) Organometallics 19:4740
81. Viau L, Lepetit C, Commenges G, Chauvin R (2001) Organometallics 20:808
82. Ohta T, Fujii T, Kurahashi N, Sasayama H, Furukawa I (1998) Sci Eng Rev Doshisha Univ 39:133
83. Spannenberg A, Baumann W, Rosenthal U (2000) Organometallics 19:3991

84. Vicente J, Chicote M-T, MacBeath C, Fernandez-Baeza J, Bautista D (1999) Organometallics 18:2677
85. Falvello LR, Fernandez S, Navarro R, Urriolabeitia EP (1996) Inorg Chem 35:3064
86. Falvello LR, Fernandez S, Navarro R, Pascual I, Urriolabeitia EP (1997) J Chem Soc Dalton Trans 763
87. Falvello LR, Fernandez S, Navarro R, Urriolabeitia EP (1997) Inorg Chem 36:1136
88. Barco IC, Falvello LR, Fernandez S, Navarro R, Urriolabeitia EP (1998) J Chem Soc Dalton Trans 699
89. Navarro R, Urriolabeitia EP (1999) J Chem Soc Dalton Trans 4111
90. Falvello LR, Fernandez S, Navarro R, Rueda A, Urriolabeitia EP (1998) Inorg Chem 37:6007
91. Falvello LR, Fernandez S, Navarro R, Rueda A, Urriolabeitia EP (1998) Organometallics 17:5887
92. Fernandez S, Navarro R, Urriolabeitia EP (2000) J Organomet Chem 602:151
93. Vicente J, Chicote M-T, Lagunas M-C (1999) Helvetica Acta 82:1202
94. Allen A, Lin Jr, Lin W (1999) Organometallics 18:2922
95. Tanase T, Ara Begum R (2001) Organometallics 20:106
96. Tanase T, Ukaji H, Takahata H, Toda T, Yamamoto Y (1998) Organometallics 17:196
97. Fernandez S, Garcia MM, Navarro R, Urriolabeitia EP (1998) J Organomet Chem 561:67
98. Larraz C, Navarro R, Urriolabeitia EP (2000) New J Chem 24:623
99. Laguna M, Cerrada E (1999) Reactivity of diauracycles. A way to prepare chains of gold atoms. In: Braustein P, Oro LA, Raithby PR (eds) Metal clusters in chemistry. Wiley-VCH, Weinheim, Germany, chap 1.24, p 459
100. Fackler JP Jr (1997) Polyhedron 16:1
101. Neitling DC, Staples RJ, Fackler JP Jr (1997) Inorg Chim Acta 263:35
102. Van Zyl WE, Lopez-de-Luzuriaga JM, Fackler JP Jr, Staples RJ (2001) Can J Chem 79:896
103. Fernandez EJ, Gimeno MC, Laguna A, Lopez-de-Luzuriaga JM, Olmos E (1998) Polyhedron 17:3919
104. Alvarez B, Fernandez EJ, Gimeno MC, Jones PG, Laguna A, Lopez-de-Luzuriaga JM (1998) Polyhedron 17:2029
105. Findeis B, Contel M, Gade LH, Laguna M, Gimeno MC, Scowen IJ, McPartlin M (1997) Inorg Chem 36:2386
106. Steiner M, Grützmacher H, Prtizkow H, Zsolnai L (1998) Chem Commun 285
107. Kalyanasundari M, Panchanatheswaran K, Robinson WT, Wen H (1995) J Organomet Chem 491:103
108. Kalyanasundari B, Panchanatheswaran K, Parthasarathi V, Robinson WT (1999) Bull Chem Soc Jpn 72:33
109. Laavanya P, Venkatasubramanian U, Panchanatheswaran K, Krause Bauer JA (2001) Chem Commun 1660
110. Falvello LR, Fernandez S, Navarro R, Urriolabeitia EP (1999) Inorg Chem 38:2455
111. Döring S, Erker G, Fröhlich R, Meyer O, Bergander K (1998) Organometallics 17:2183
112. Fritz G, Scheer P (2000) Chem Rev 100:3341
113. Korth K, Schorm A, Sundermeyer J, Hermann H, Boche G (2000) Iodotrimethylsilylmethylene-triphenylphosphane. In: Weis J (ed) Organosilicon chemistry IV: from molecules to materials. Wiley-VCH, p 238
114. Gilheany DG (1994) Chem Rev 94:1339
115. Grumbine SK, Mitchell GP, Straus DA, Tilley TD, Rheingold AL (1998) Organometallics 17:5607
116. Schrödel HP, Schmidpeter A, Nöth H (1998) Z Naturforsch 53b:1285
117. Schrödel HP, Jochem G, Schmidpeter A, Nöth H (1995) Angew Chem Int Ed Engl 34:1853
118. Breitsameter F, Schmidpeter A, Nöth H (2000) Chem Eur J 6:3531
119. Schmidpeter A, Jochem G, Klinger C, Robl C, Nöth H (1997) J Organomet Chem 529:87
120. Breitsameter F, Schrödel HP, Schmidpeter A, Nöth H, Rojas-Lima S (1999) Z Anorg Allg Chem 625:1293
121. Breitsameter F, Schmidpeter A, Nöth H, Knizek J (1999) Z Naturforsch 54b:1

122. Breitsameter F, Mayer P, Schmidpeter A (2000) Z Naturforsch 55b:519
123. Breitsameter F, Schmidpeter A, Schier A (1998) Eur J Inorg Chem 381
124. Breitsameter F, Polborn K, Schmidpeter A (1998) Eur J Inorg Chem 1907
125. Il'in EG, Kovalev VV, Aladzheva IM, Bykhovskaya OV, Leont'eva IV, Mastryukova TA, Buslaev YA (1998) Russ J Gen Chem 68:1378
126. Smolli OB, Panchishin SY, Romanenko EA, Drach BS (1999) Russ J Gen Chem 69:1583
127. Hamzaoui M, Provot O, Camuzat-Dedenis B, Moskowitz H, Mayrargue J, Cicéron L, Gay F (1998) Tetrahedron Lett 39:4029
128. Camuzat-Dedenis B, Provot O, Moskowitz H, Mayrargue J (1999) Synthesis 1558
129. Aitken RA, Drysdale MJ, Ferguson G, Lough AJ (1998) J Chem Soc Perkin Trans 1 875
130. Aitken RA, Drysdale MJ, Ryan BM (1998) J Chem Soc, Perkin Trans 1 3345
131. Aitken RA, Armstrong JM, Drysdale MJ, Ross FC, Ryan BM (1999) J Chem Soc Perkin Trans 1 593
132. Chen X-D, Liu Y-X, Cao R-Z, Xue M, Liu L-Z (1999) 152:185
133. Abdou WM (1999) Phosphorus Sulfur Silicon (1999) 144/146:393
134. Silveira CC, Perin G, Boeck P, Braga AL, Petragnani N (1999) J Organomet Chem 584:44

Top Curr Chem (2003) 229: 75–94
DOI 10.1007/b11151

Donor-Acceptor Complexes of Low-Coordinated Cationic π-Bonded Phosphorus Systems

Wolfgang W. Schoeller

Fakultät für Chemie der Universität, Postfach 10 01 31, 33501 Bielefeld, Germany
E-mail: wolfgang.schoeller@uni-bielefeld.de

Cationic low-coordinated π-bonded phosphorus compounds add Lewis donors, such as amines or phosphines. In contrast to the trigonal bipyramide formed in donor-addition to the siliconium cation, the phosphorus cations add donors in a perpendicular fashion, depending on the nature of the π-bonds toward phosphorus. According to quantum chemical calculations the various cations reveal different stabilities and hence a strong variation in donor addition abilities.

Keywords. Cations, π-Bonded, Low-coordinated phosphorus systems, Donor-addition, Mono- and bis-coordination, Bipyramide formation

1 Introduction, the Nature of the Donor-Acceptor Complex

Donor-acceptor compounds have been known for a long time [1] and their understanding has existed long on the basis of the acid-basis theory [2]. The appearance of a standard textbook in 1961 by Briegleb [3] had a great impact in this field, since it was a comprehensive review at this time. The modern understanding of donor-acceptor complex formation [4] or more general the nature of donor-acceptor bonding was initiated by Mulliken in a series of publications [5–7]. Notable contributions to the theory of molecular complexes have also been made by Weiss [8], Brackmann [9] and Dewar [10]. Several reviews on the subject of donor-acceptor complexes have been published, the most recent one is the late contribution of Haaland [11].

An exemplary view of a donor-acceptor complex is given in **1** (Scheme 1).

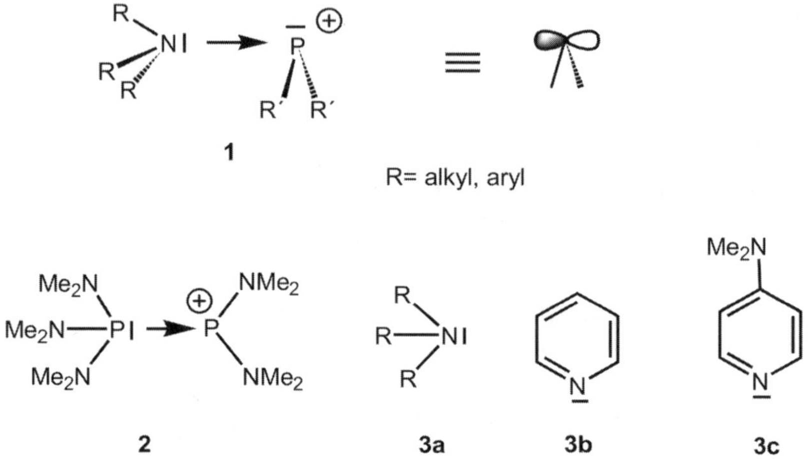

Scheme 1

Electron density is transferred from a Lewis donor to a Lewis acceptor, here shown for an amine (donor) interacting with a phosphenium cation (acceptor). The lone pair orbital at the amine dives into the empty *p*-orbital of the phosphenium cation. Following the ideas of Mulliken the donation towards an acceptor can be viewed as sketched in Scheme 2.

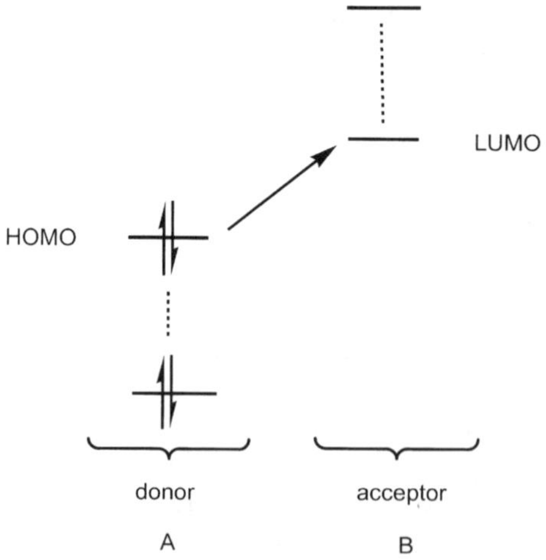

Scheme 2

The donor-acceptor formation can be considered by transfer of electrons from the donor to the acceptor. In principle one can assume donor-acceptor interaction from A (donor) to B (acceptor) or alternatively, since B (A) has also occupied (unoccupied) orbitals, the opposite charge transfer, from B to A. Such a view refers to mutual electron transfer and has been commonly established for the analysis of charge transfer spectra of π-complexes [12]. A classical example for a donor-acceptor complex, **2**, involving a cationic phosphorus species has been reported by Parry et al. [13]. It is considered that the triaminophosphines act as donor as well as an acceptor towards the phosphenium cation. While **2** refers to a P-donor, *N*-donors are in general more common, as for example amines, **3a**, pyridines, **3b**, or the very nucleophilic dimethylaminopyridine (DMAP) [14], **3c**. It is even a strong donor towards phosphorus trichloride [15].

In more detail, the interaction energy between donor and acceptor is determined by the ionisation potential of the donor and the electron affinity of the acceptor. The interaction energy increases with lowering of the former and raising of the latter. In the Mulliken picture (Scheme 2) it refers to a raising of the HOMO (highest occupied molecular orbital) and lowering of the LUMO (lowest unoccupied molecular orbital). Alternatively to this picture donor-acceptor formation can be viewed in a Born-Haber cycle, within two different steps (Scheme 3).

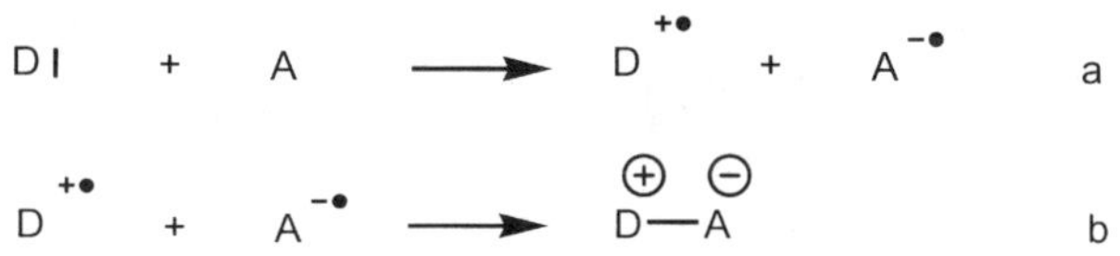

Scheme 3

First one electron is transferred from the donor to the acceptor (Scheme 3, **a**) and then both fragments form a covalent, but polar two-electron two-centre bond (Scheme 3, **b**) [16]. As we will show in further discussion this view has some advantages over the formulation of a donor-acceptor as a charge transfer complex, as has been originally viewed by Mulliken [5] and Dewar and Lepley [12]. The step **a** is a one-electron redox process and is determined by the ionisation potential of the donor and the electron affinity of the acceptor, i.e. intrinsic properties of both fragments. The larger the electron affinity – as a rule cations possess much larger electron affinities than neutral Lewis acceptors – the more important are these considerations. In the final step **b** the covalent bond is formed. This can be attributed to the bond energy which would result from the recombination of two radicals plus the electrostatic energy between a positively charged D and a negatively charged A.

In practice one can differentiate between two kinds of donors, the resulting donor-acceptor bonds can be largely ionic (class I) or covalent (class II) [16]. Donors of the first type (class I) stem from the first row of the periodic table of elements, such as amines, ethers, in detail structures in which the Lewis basis centre possessing the non-bonding lone pair is strong electronegative. Donors of the second type are constituted from elements of the second row of the periodic table of elements, such as phosphines, thioethers, etc. (class II). These Lewis donors are less electronegative compared with their first row counterparts but tend to form stronger covalent donor-acceptor bonds. These bonding facettes are emerging mainly from theoretical considerations [16]. In most experimentally reported cases donors are restricted to *N*-donation (amines, pyridines, etc.) towards electron accepting precursors. The only experimental detailed study in which different cases of donors (*N*-donors, *O*-donors, etc.) were probed, was made only recently [17] for the case of formation of trigonal -bipyramidal coordinated siliconium ions, **4** (Scheme 4).

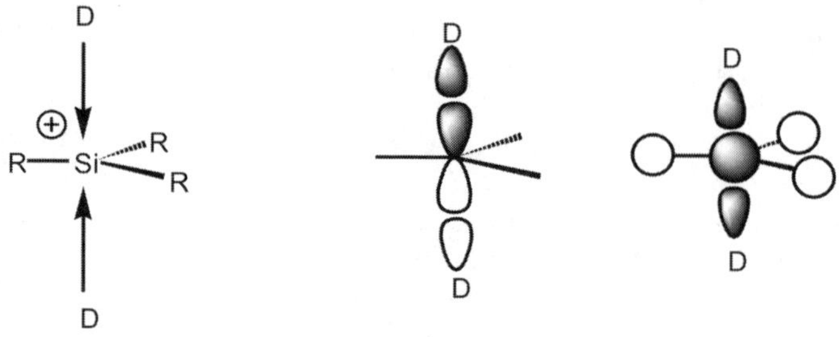

4 **A** **B**

D = Lewis donor

Scheme 4

Even for this case most of the experimental reports refer to *N*-donation at the siliconium ion [18, 19]. It must be noted that organosiliconium ions are very strong acceptors and can be stabilized by various donors [20], σ-donors [21, 22] and π-donors [23–25]. The strong electrophilicity of the siliconium ion [26] as well as *N*-donation [27] were studied also in detail by quantum chemical considerations. The systematic study of the addition of the various donors towards Lewis acid cations in phosphorus chemistry is in fact lacking. Since the electron affinities of cationic species are much larger than those of neutral Lewis acids, one expects that overall the resulting donor-acceptor complexes with cationic species are essentially more strongly bound as compared with neutral Lewis acids.

Bis-coordination of two donors towards the silyl cation is determined by the formation of two axial bonds, with formation of predominantly *p*-electron containing axial bonds, **A**, as emphasized in the Musher concept [28] and additional *s*-contribution, **B**. This is the consequence of the tendency of the higher main-group element silicon towards orbital non-hybridization [29–31] and induces a hydridic character of the hydrogen (at Si) by coordination of the donors at the electrophilic siliconium ion $R_2Si^{(+)}H$. All these aspects are well characterized for the siliconium ion and await corresponding characterization for the low-coordinated phosphorus cations. The stability of the two axial bonds is depending on the electronegativities of the donor-centres and the accepting centre. It increases with increasing electronegativity of the former and decreasing electronegativity of the latter.

2 Cationic π-Bonded Species

2.1 Theoretical Consideration of Cation Stabilities and Electronic Structures

It is informative to begin the further discussion with a consideration of the various cations which will be the aim of the following sections. The most prominent examples of the low-coordinated cations in phosphorus chemistry are the phosphenium (phosphanylium) cations. They were scrutinized in recent reviews [32–35], and hence will not be discussed in details in the present review.

For the π-bonded cationic species under discussion it is of importance to have knowledge about (a) the intrinsic stability of the cation itself and (b) the electronic nature of the frontier orbitals. The intrinsic stability of the cation is determined by the extent of delocalisation of the positive charge over the entire structure and the frontier orbitals should give some information on its reactivity towards nucleophiles (Lewis bases). The donors should approach the positions of the highest electrophilicity (energy lowest LUMO with the highest orbital coefficient at the approaching atom) within a given structure. An attempt to rationalize various cationic structures in low-coordinated phosphorus chemistry has been provided on the basis of quantum chemical calculations [36] and has been extended lately to cyclic π-systems of the Arduengo-type [37] (Fig. 1).

A quantitative measure of the stabilities of cations **5–11** is provided by reaction **12** (Fig. 1, bottom). It sets the intrinsic cation stabilities in reference to the

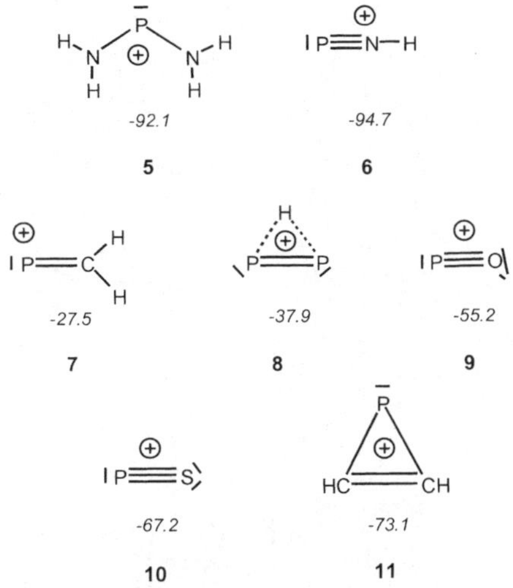

$$HPR^1R^2 + PH_2^{(+)} \longrightarrow PR^1R^{2(+)} + HPH_2 + \Delta E \qquad (12)$$

Fig. 1. Relative stabilities (in italic, kcal/mol) of π-bonded cations, as obtained from quantum chemical calculations, according to reaction **12**

unsubstituted phosphenium cation $H_2P^{(+)}$. Although these data were computed for the gas phase they nevertheless yield the first-order effects of the cation stabilities. In practice they also help one to understand the different reactivity patterns of the various cations.

According to these consideration the diamino-substituted phosphenium (an alternative suggestion for its nomenclature is phosphanylium) cation, **5**, and the phosphanetriylammonium (iminophosphenium) cation, **6**, possess the largest intrinsic (gas phase) stabilities. Since in the X-ray structures the molecules are to a first-order isolated, this theoretical stability scale determined for the gas phase should also mimic the various trends of the stabilities of the cations and their chelation behaviour. The methylenephosphenium, **7**, and the P_2H^+ cations, **8**, suffer from poor stabilities. On the other hand the phosphirenium cation, **11**, is considered to be fairly well stabilized. It is due to π-electron delocalisation of the positive charge in the phosphirenium cation. Intermediate cases in stability are the PO^+ (**9**) and PS^+ cations (**10**). Of further interest are the frontier orbital considerations, as shown in Fig. 2.

Accordingly, various classes of cations can be differentiated. The cations **5** and **11** are carbene types (Fig. 2A), in other words they possess a σ-orbital (filled with

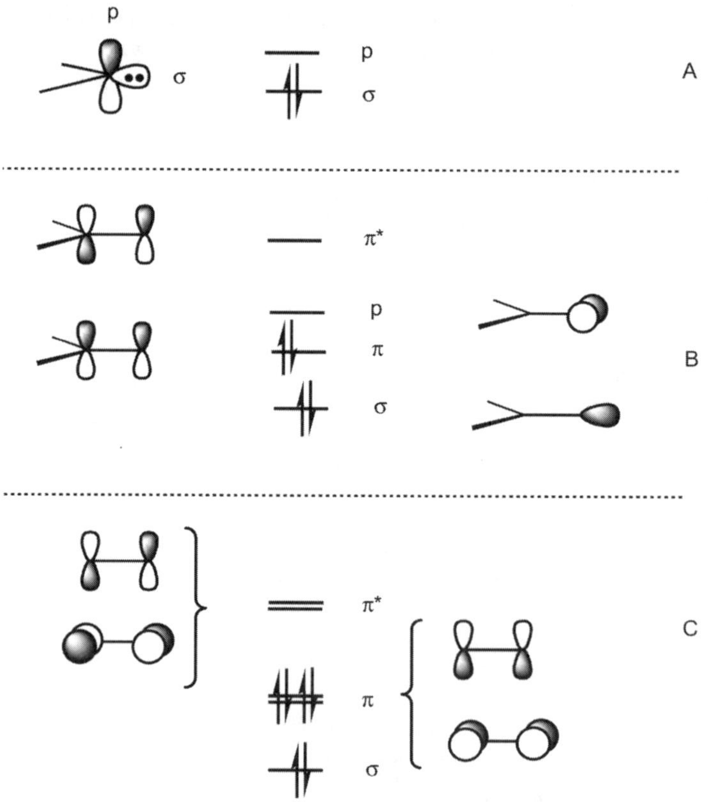

Fig. 2. Frontier orbitals of different types of cations

two electrons) and an empty *p*-orbital, perpendicular to it. The singlet character of these species is ensured by the considerable energy separation between these two frontier orbitals. A nucleophile (or Lewis base) should add perpendicular to the plane spanned with the phosphorus atom. Possible dicoordination of a donor is expected to occur in a bis-axial orientation of the two donors. In contrast, **7** possesses an empty *p*-orbital in plane with the PCH_2 unit, (Fig. 2B). Again a different situation (Fig. 2C) is provided in **6**, **9** and **10**. These cations bear a degenerate set of π-, π*-orbitals, which gives rise to bis-donor-addition in an orthogonal fashion. In other words the first donor adds to one polarized π-bond and the second one to its orthogonal counterpart. Hence, both donors are not axially oriented, as is well known for the arrangement of the trigonal bipyramide (e.g. in the siliconium cation); rather they span an angle ∠DPD (D=donor) of approximately 90°. These considerations have to be regarded as rules for donor-addition to these species. As will be shown by the experimental structural data for **6**, these principal considerations which emerged from the quantum chemical calculations are followed. More details on these aspects (charge densities, etc., on the investigated cations) can be found in the cited paper [36]. A particular case is given for **8**; in its free form it is predicted to adopt a bridged or a strongly bent structure, depending on the substituent attached at phosphorus (see below).

In the first part of this review we report on the species of the type PE^+, **6**–**10** (Scheme 5).

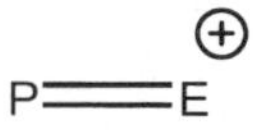

E: **6** NR; **7** CR_2; **8** PR; **9** O; **10** S.

Scheme 5

2.2
Cationic Species of the Type PNR^+, Mono-Coordination

The best investigated cationic species is the phosphanetriylammonium system, **6** (E=NR). The electrophilic attack of $AlCl_3$ on the P=N double bond [38, 39] of the aminoiminophosphane R_2N-P=NR (R=$SiMe_3$) generated the internal salt R_2N-$P^{(+)}$-N($AlCl_3^{(-)}$)R. Since that time the reaction of iminophosphanes with Lewis acids has being considered as a promising synthetic pathway to phosphanylium species [34]. Furthermore, it was reported [40] that the reaction of the chloroiminophosphane ClP=NMes* (here and further Mes*=2,4,6-tris(t-Bu)-C_6H_2) with $AlCl_3$ leads to a stable $AlCl_4^-$ adduct, instead of the expected phosphanylium cation (Scheme 6).

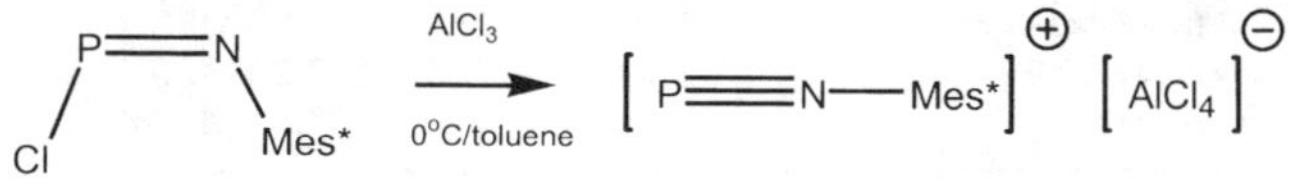

6a

Scheme 6

The solid structure of **6a** indicates the presence of an ion pair, the cation molecule is only weakly surrounded by two $AlCl_4^-$ units. The $PNC^{ipso}(Ar)$ angle (177°) refers to the almost ideal sp-hybridization at the nitrogen. The fairly short PN bond (1.475 Å) indicates a high degree of triple-bond character and is in agreement with quantum chemical calculations [41] (1.476 Å). Consequently **6a** can be considered as isolobal to the diazonium salts of the type $[Ar–N{\equiv}N]^+$ $[X]^-$ or to phosphalkynes [42]. The cation **6a** possesses positive charge mainly localized at phosphorus and an energetically low-lying set of π* orbitals.

As has been shown the structure and reactivity of the iminophosphanes of the general type X–P=*N*–Mes* are considerably dependent on the substituent X bonded to the two-coordinate phosphorus [43, 44]. According to the size of the dialkylamino group in aminoiminophosphanes (R_2N)–P=NMes*, cis (*Z*) **13a** or trans (*E*) isomers **13b** are stable in the solid state (Scheme 7). In a cis configuration the *s*-character of the phosphorus-nitrogen multiple bond is more pronounced than in the trans configuration. Within the series compounds RO-P=NMes*, **14** (Fig. 3), the electronic nature of the substituent R exerts a shortening of the PN bond with concomitant lengthening of the PO bond. In addition the angle ∠PNR increases. The increase in the electronegativity of the RO group leads from structures with a covalent P-O bond to those where the $[RO]^-$ anion is separated from the cationic fragment $[PNMes^*]^+$ [45].

These effects are also accompanied by a shielding of the ^{31}P nuclei. These data lucidly describe the change of the iminophosphane to the cation; the results of

13a **13b**

Scheme 7

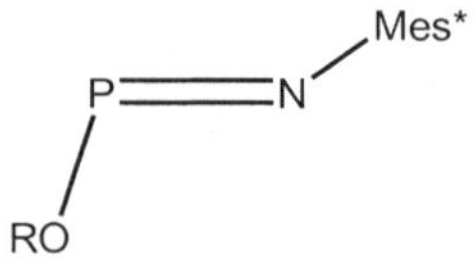

14

R	PN	PO	∠ OPN	∠ PNC
Me_3Si	1.529	1.585	115.8	144.4
*t*Bu	1.528	1.601	110.0	144.7
tBu_3C	1.487	1.615	110.0	175.3
o-Me-C_6H_4	1.497	1.641	111.8	164.1
2,6-*t*Bu-4-Me-C_6H_2	1.500	1.658	110.3	173.7
p-Me-C_6H_4-SO_2	1.492	1.728	107.2	150.1
CF_3-SO_2	1.467	1.923	108.4	176.4

Fig. 3. Selected bonding parameters (bond lengths in Angströms, bond angles in degrees) of substituted iminophosphanes, 14

another investigations were reported [46] which confirm the previous suggestion. Similar effects are observed for iminophosphanes of the type C_5Me_5-P=NR, **15**, with increasing donor ability of the nitrogen [47,48]. It results in a π-complex between a cyclopentadienyl anion and an imino-phosphenium cation (*E*) (Scheme 8). The distances between the phosphorus atom and the centre of the aromatic ligand results to 2.69 and 3.03 Å and decreases with increasing π-donor contribution of the aromatic ligand R (at N). Besides the $Cp^{(-)}$-coordination just mentioned, the iminophosphenium cation forms π-complexes with the neutral arenes [49], **16**. Thus this donor-adduct can be related to **15**.

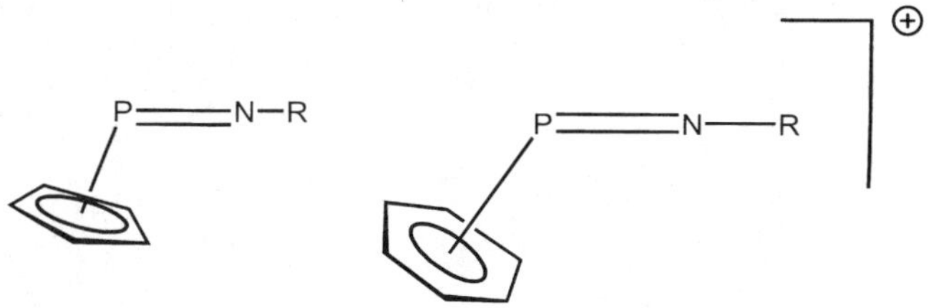

15 16

Scheme 8

While these reports are mainly concerned with an evaluation of the peculiar structure of the iminophosphenium cation, this species also reveals cycloaddition properties. The cation shows a [2+1] cycloaddition behaviour towards alkynes [50]. With iminophosphanes and alkylazides the cations react with the formation of four- and five-membered heterocyclic ring systems [51], as shown for example in Scheme 9.

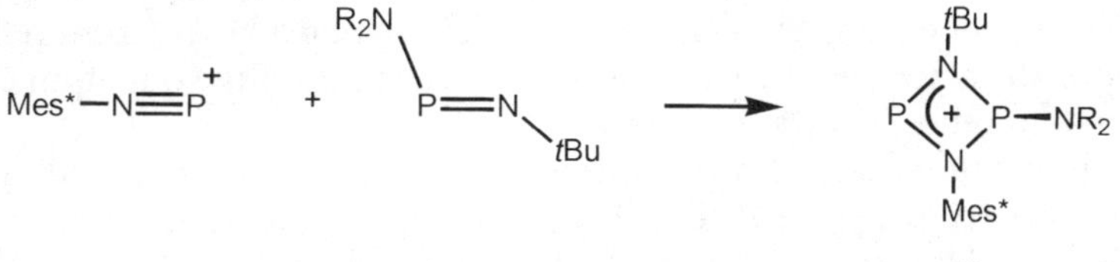

17

Scheme 9

The product **17** of the [2+2] cycloaddition reaction can be alternatively considered as an intramolecular donor-acceptor complex.

The iminophosphenium cation was also of interest for more physical investigations. It was noted to be a stable entity in gas phase experiments [52], the ^{31}P parameters were investigated in detail [53] and a systematic study of the nucleophilic addition of CH, NH and OH bonds was performed [54] with a concomitant interpretation of the ^{31}P chemical shifts (at the dicoordinate phosphorus centres). The latter authors also confirmed the loose interaction of a triphenylphosphine with the iminophosphenium cation (PP=2.625 Å).

2.3
Cationic Species PNR$^+$, Dicoordination

The first product of dicoordination, **18,** was synthesized via treatment of phosphinoiminophosphanes with selenium. The latter then inserted into the PP-bond and formed finally a bis-coordinated donor-acceptor complex [55]. The overall synthetic route is shown in Scheme 10.

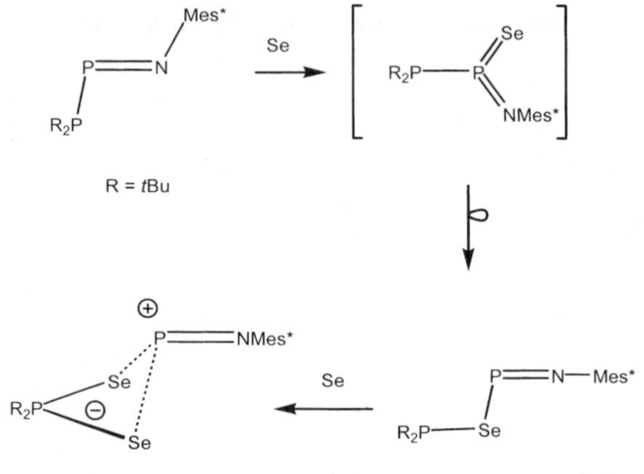

Scheme 10

The two selenium atoms are attached to the low-coordinated phosphorus atom, with two lengthened dative bonds (2.788 and 2.637 Å). Interestingly the phosphorus atom is here strongly pyramidalized, the ∠ SePSe angle was 80.6°. In other words the lone pair orbital at the electrophilic phosphorus atom remains stereochemically active, as one would expect on the basis of the quantum chemical theory of bonding within this species.

Reactions of the iminophosphenium cation tosylate with an equimolar amount of quinuclidine or 2,2′-dipyridyl proceed almost instantaneously and quantitatively to a bis-coordinated adduct **19** [56] (Scheme 11).

The N → P dative bonds are weak and different in lengths (1.800 Å on average), and the triflate anions are effectively extended to consider interaction with the counter ion. Again the phosphorus atom is strongly pyramidalized and features the aspects of an inert nonbonding electron pair.

The product of bis-coordination, **20** at phosphorus, could be also achieved utilizing the more nucleophilic *p*-dimethylamino pyridine (DMAP) [57, 58]. As shown in Scheme 12, bis-coordination of the two donors occurs in general by taking up the two orthogonal positions of the PN π-bonds [58].

Similar structural results are obtained if one donor is neutral (e.g. DMAP) and the other donor is anionic (Cl$^-$) (structure **21**). The adducts possess extremely soft potentials and in solution the ^{31}P NMR shifts are strongly temperature dependent [59]. It confirms the assertion that the adducts are very labile.

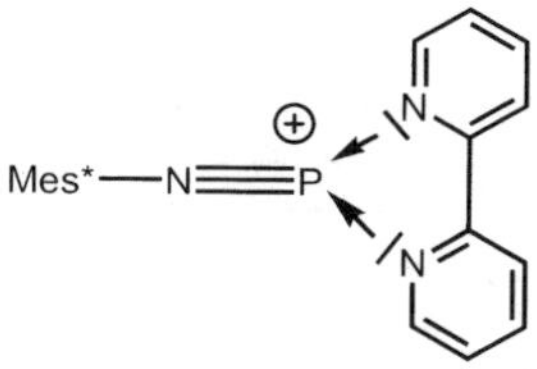

19

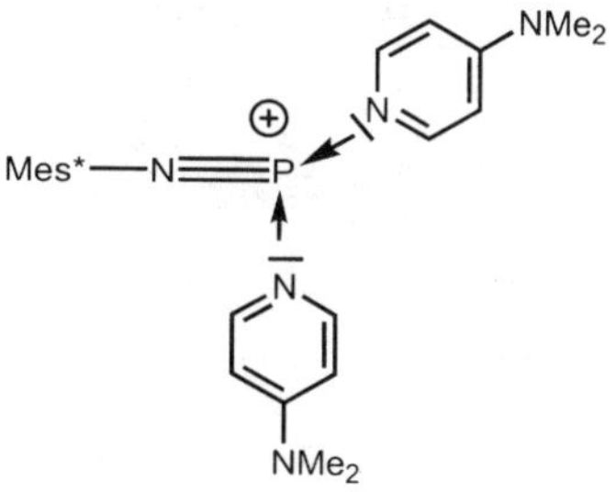

20

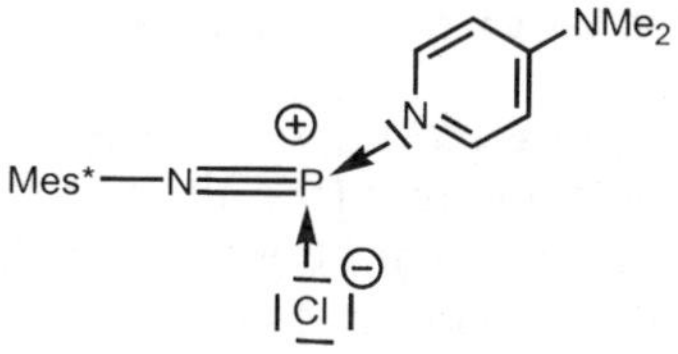

21

Scheme 11

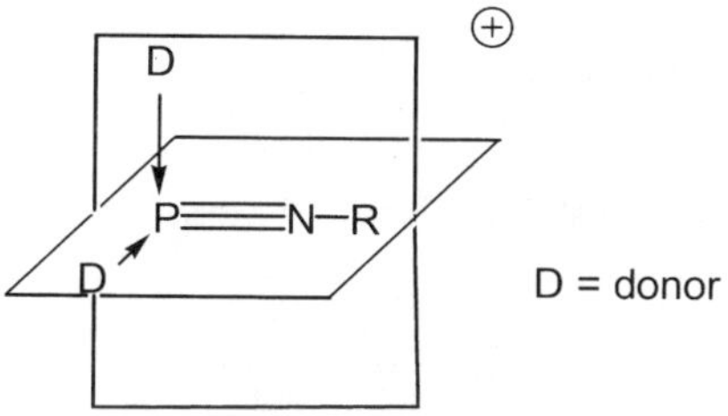

Scheme 12

Overall the quantum chemical calculations predict that one has to differentiate between three different types of donors adding to the iminophosphenium cation, types A to C (Scheme 13), which can be formally attributed to bis-donor complexes. In type A and B the two donors add almost orthogonal to each other and the bonding facettes of the PN moiety (a short PN distance and an almost linear arrangement with the adjacent ligand) are retained. Alternatively when an anionic donor adds, type C, the PN bond is lengthened, the PNC^{ipso}(Ar) increases, and negative charge builds up at the imino-nitrogen.

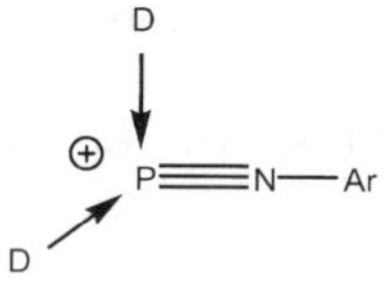

A

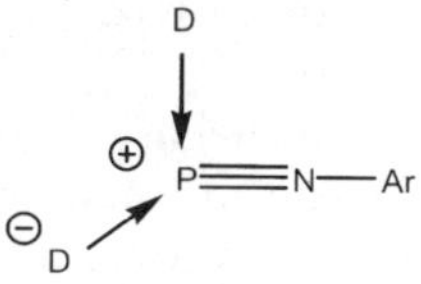

B

Scheme 13 a, b

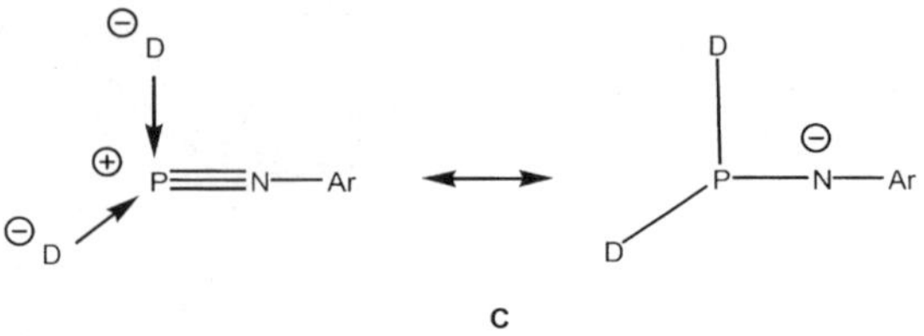

Scheme 13 c

2.4 The Phosphanetriylphosphonium Cation, Mono-Coordination

In contrast to the iminophosphenium cation, **6**, described in the previous section, the corresponding phosphaphosphenium (the alternative nomenclature to this species is phosphanediylphosphenium or phosphanetriylphosphonium), cation **8**, shows only a modest intrinsic stability, as suggested by the quantum chemical calculations (see Fig. 1). They predict that a protonated P_2 should have a bridged structure [60–66], **8a** (Scheme 14).

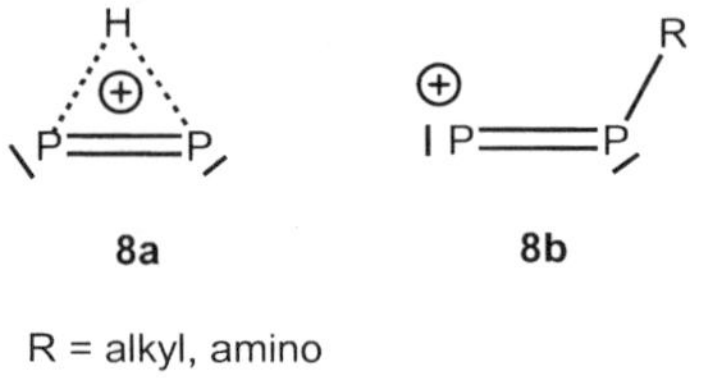

Scheme 14

On the other hand the alkyl- or amino-substituted congeners should adopt a more classical structure, in which the twofold coordinated phosphorus atom is bent, **8b**. A first hint of the cation was reported by mass-spectroscopic investigations [67]. The synthetic verification of this prediction starts from amino-substituted diphosphene **22** via protonation with CF_3SO_3H [68] (Scheme 15).

In the structure **24a**, the triphenylphosphine is strongly bound to the electrophilic phosphorus centre (PP=2.206 Å) which indicates a strong covalent character of this bond. Upon warming the solution to 20 °C decomposition takes place and a mixture of bicyclotetraphosphanes is formed. Interestingly, some structural trends towards the formation of ion pairs between a donor and an acceptor were also reported in the „push-pull" diphosphene structures **25**–**27** [69] (Fig. 4).

With increasing electronegativity of one ligand in the diphosphene the ∠PPR angle shrunk and decreased finally below 90° for the diphosphene **27**. The experimental findings were accompanied by the quantum chemical calculations. There are no other reports regarding further experimental investigations or possible dicoordination in these compounds.

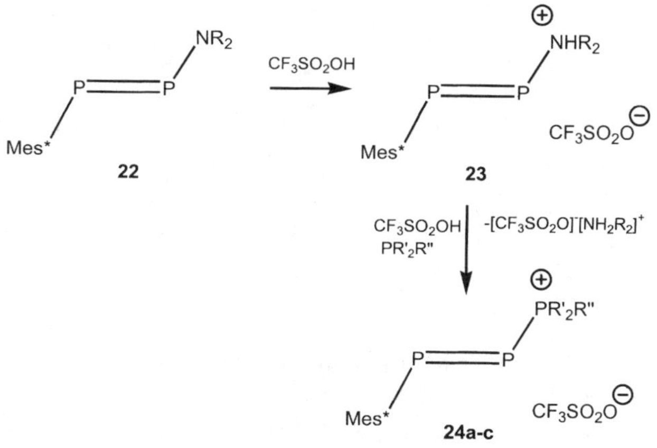

R_2N = iPr_2N; 2,2,6,6-tetramethylpiperidine (TMP); cyclohexylN

24: a R'=R''=Ph; **b** R'= t-Bu, R''= H; **c** R'= H R''=Mes*

Scheme 15

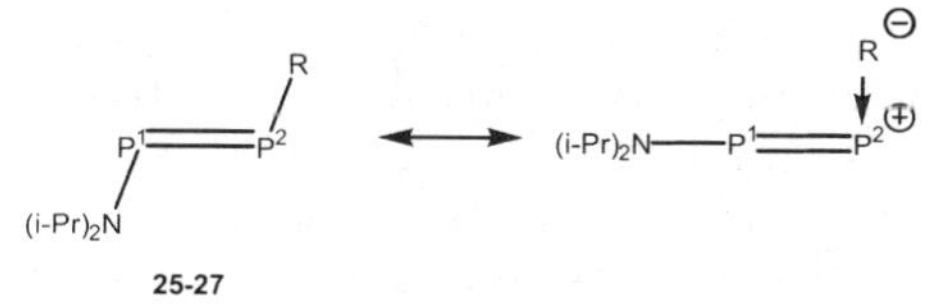

Compound	R	P^1P^2	$\angle NP^1P^2$	$\angle RP^2P^1$
25	Mes*	2.018	101.2	92.3
26	$(SiMe_2tBu)_2$	2.011	110.5	89.8
27	TMP	2.029	114.8	89.4

Fig. 4. Push-pull substituted diphosphanes, bonding parameters in Angströms and degrees

The investigations indicate that, in agreement with the calculations on the cations, the push-pull substituted diphosphenes tend to form a bridged structure of one ligand; the other substituent can easily depart under formation of an ion pair structure.

2.5
The Methylenephosphenium Cation

A phosphine-donor stabilized adduct **29** of the methylenephosphenium ion, **7**, was synthesized [70] in a way similar to the iminophosphenium cation **6** from the corresponding chloroderivative **28** and PPh_3 with addition of $AlCl_3$. Interestingly the base-free methylenephosphenium cation has recently been observed in the gas phase [71] (Scheme 16).

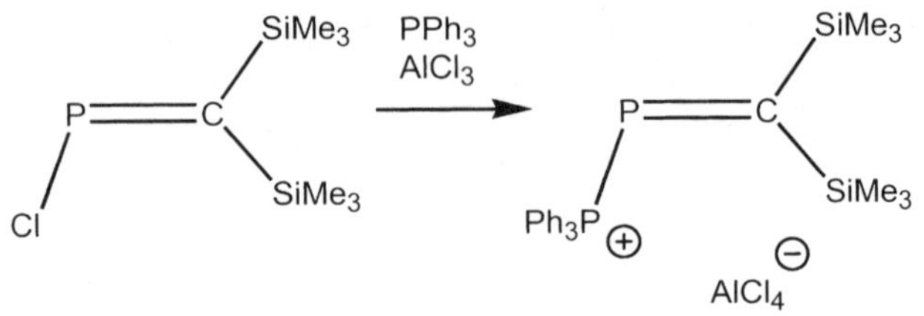

28 29

Scheme 16

The unequivocal support for the constitution as an ion pair came from the structure analysis. It indicated the formation of an isolated anion $[AlCl_4]^-$; interestingly the shortest Cl(1)-P(1) contact (3.69 Å) falled outside the sum of the van der Waals radii. However the PP bond to the phosphine donor was only slightly longer (2.27 Å), as compared with other related species, such as the phosphino-substituted methylenephosphane. An alternative species **29a** was described with a triflate as anion. While the former remained unchanged at room temperature the latter decomposed quantitatively to the phosphaalkyne $P{\equiv}C{-}SiMe_3$ (**30**) by loss of PPh_3 and $CF_3SO_3SiMe_3$ (Scheme 17). The reaction conditions differ markedly from the generally very drastic reaction conditions required in the synthesis of phosphaalkynes (via an elimination reaction) [72]. From the donor-adduct a cationic Ni-metal complex could also be synthesized. The donor-stabilized cation offers a synthetic route for mild synthesis of hitherto unknown phosphaalkynes. Amine coordination has a stronger catalytic effect on this reaction than phosphine coordination [73].

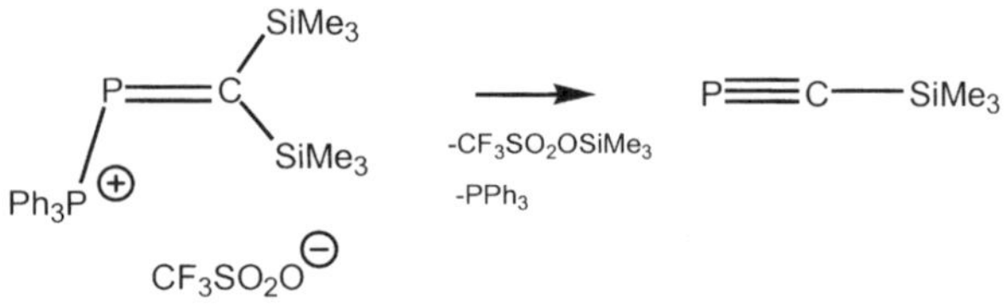

29a 30

Scheme 17

Another approach to a donor adduct of the methylene phosphenium cation is the addition of a phosphonium cation to the phosphaalkyne. The reaction of the protic cation $[HPPh_3]^+[CF_3SO_3]^-$ with $C_{10}H_{15}CP$ produced a white powder which was identified as the P-phosphonio-substituted phosphaalkene [74]. Alternatively to the elimination reaction the phosphaalkynes were protonated. C-protonation of adamantylphosphaacetylene and *tert*-butylphosphaacetylene occurred in superacid media under formation of phosphavinyl cations. From these spirocyclic betaines by reaction of 1-Ad-C≡P (Ad = adamantyl) with $B(OTf)_3$ a phosphavinyl cation could be detected [75].

Overall, for the free methylenephosphenium cations quantum chemical calculations predict particular bonding properties, depending on the nature of the substituents attached at the carbon position [76] (Scheme 18). With R=phenyl (**7a**), phosphino (**7c**) or amino group (**7b**) as substituent a bridged structure is predicted while by the silyl substitution at carbon (**7d**) the C-Si bond is considerably lengthened by hyperconjugative interaction with the neighbouring positively charged phosphorus atom.

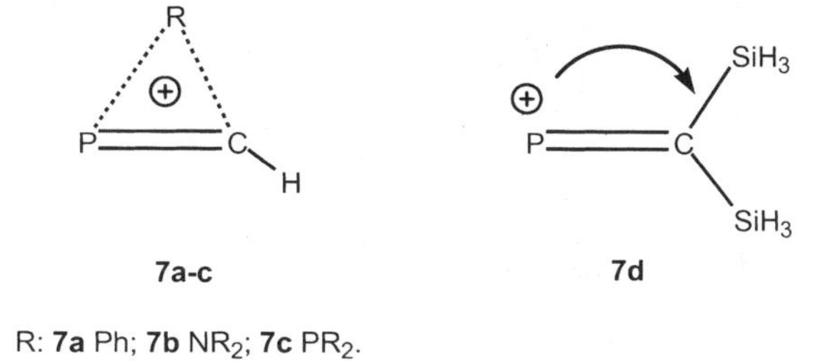

Scheme 18

2.6 The Phosphirenylium Cation

In contrast to the cyclopropenylium ion, which has been known for a long time, its phosphorus analogue was first synthesized just recently. The phosphirenylium cation, **32**, can be obtained by addition of the strong Lewis acid $B(OTf)_3$ to a solution of the 1*H*-phosphirene, **31**, in liquid SO_2 [77] (Scheme 19).

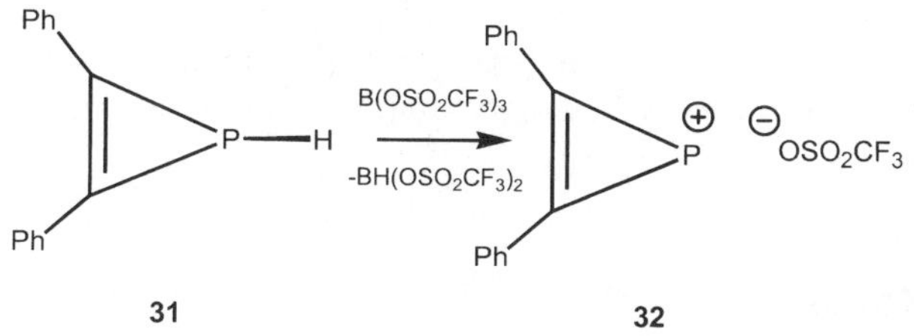

Scheme 19

It is the only example of a free, persistent phosphirenylium ion, and also, only one stable transition-metal complex of this species was published [78, 79]. Quantum chemical calculations [80, 81] indicated that in the halogeno-phosphirenes the P-X bonds already possesses a high ionic character and can be described as interactions between phosphirenylium and halide ions. The aromatic character of the phosphirenylium ion was shown to be based on a three-centre two-electron bond of π-type and the resonance energy was assessed by calculation to 38 kcal/mol. Before the generation of **32**, substituted phosphirenylium ions were noticed in the mass spectra of 1*H*-phosphirenes [77] and the parent ion was found in an ion gas-flow reaction of PH_n^+ and P_n^+ cluster ions in the mass spectrometer [82, 83]. From the calculations it has been stated that amino substitution should stabilize the phosphirenylium ion and hence reduce the strength of the P-X bond. Furthermore the cation is stabilized by complexation with one or two SO_2 molecules. While the phosphirenylium cation has been studied in detail in superacid media as well as by quantum chemical calculations, no further reports are recorded on the formation of donor-acceptor complexes.

It must be noted that the saturated ring system, the 1-chloro-λ^3-phosphirane was synthesized [84]; a cyclic phosphenium cation was postulated which is stable towards ring opening [85]. A corresponding cation could not be isolated, due to a methanid shift from a trimethylsilyl substituent.

2.7
Protonation of the Tetraphosphacubane

Photoelectron spectroscopy, ab initio calculations [86] and NMR spectroscopy [87] support the contention that in the tetraphosphacubane, **33**, n P$\rightarrow\sigma$ P-C delocalisation occurs (Scheme 20).

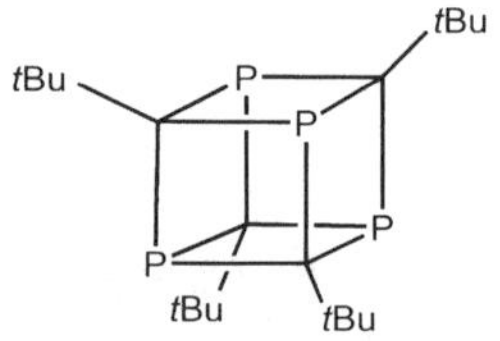

33

Scheme 20

Ethyl triflate and benzyl triflate react with the cubane at room temperature and yield phosphonium salts. Upon protonation even di- and trications could be observed [88, 89].

3
Allylic Type Cations

According to quantum chemical calculations the cations PO_2^+ and PS_2^+ should be linear [90, 91] while in the corresponding 2-phosphallyl cation the central phosphorus atom is reluctant to sp-hybridisation [92] and is bent. Bis-donor adducts **34**–**36** were reported for a variety of these cations with dimethylaminopyridine (DMAP) (Scheme 21).

The investigations included donor adducts with $PO_2^{(+)}$ [93], **34**, $PS_2^{(+)}$ [94], **35**, and $P(NMes^*)_2^{(+)}$ [95], **36**. While **34** was only formulated as an intermediate species, the other donor-acceptor complexes, **35** and **36**, were characterized by X-ray investigations. To complete this series it may also be compared with bis(ylide)-substituted phosphonium halides [96], **37**. For these cases the donors refer to

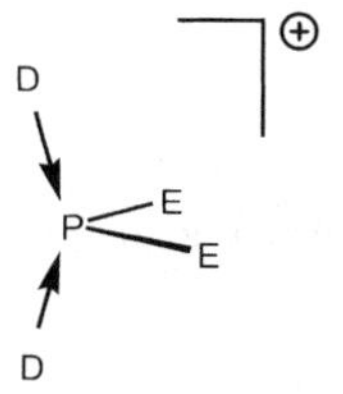

34-36

E: **34** O; **35** S; **36** NMes*; D= p-dimethylamino pyridine

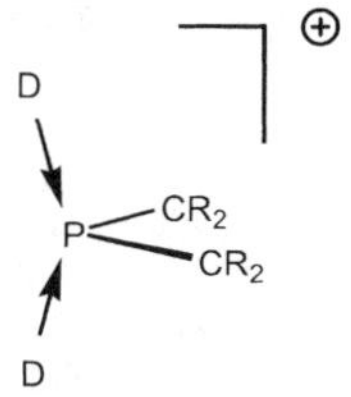

37a,b

D: **a** Cl^-; **b** Br^-

Scheme 21

halogen anions (D=$Cl^{(-)}$, $Br^{(-)}$). If one considers the P-Cl bond as very polar **37a** can be alternatively viewed as a bis-donor acceptor complex of the 2-phosphaallyl cation with two Cl^- anions. These donor-adducts have a remarkable characteristic. The plane DPD (D=donor) is not perpendicular to the plane spanned by the allyl moiety. It is the consequence of a weak Jahn-Teller distortion, operating in these structures, as suggested by quantum chemical calculations [97].

4
Conclusions

In this progress report we have reviewed the latest developments in the large area of cationic low-coordinated species and their coordination with Lewis donors. It is clear that these species are of a broad interest, in particular for catalysis. In some cases, e.g. the methylene phosphenium cation, the donor adducts also open new routes for synthesis. Regarding the mechanism for the diverse donor-addition reactions, the structural details are only poorly understood and need a better classification. In particular the variation of the Lewis-donor has to be established. Hitherto in most cases *N*-donation is studied. It includes amines or pyridines. Obviously the effect of other donors, such as phosphines, thioethers needs to be studied as well. The siliconium cation for which these effects are better known could provide an understanding for further investigations within this field.

Acknowledgements. This work has been supported by the Fonds der Chemischen Industrie and the Deutsche Forschungsgemeinschaft. I also acknowledge here the help in the work on this manuscript by Dr. Alexander B. Rozhenko. Most important is also the cooperation with Prof. Niecke over many years, whom I would like to thank.

5 References

1. Pfeiffer G (1994) Organische Molekülverbindungen, 2nd edn. F. Enke, Stuttgart, p 1927
2. Lewis GN (1938) J Franklin Inst 226:293
3. Briegleb G, Czeballa J (1960) Angew Chem 72:401; Briegleb G (1961) Elektronen-Donator-Acceptor-Komplexe, Springer, Berlin Heidelberg New York
4. Benesi HA, Hildebrand JH (1949) J Am Chem Soc 71:2703
5. Mulliken RS (1950) J Am Chem Soc 72:600
6. Mulliken RS (1952) J Am Chem Soc 74:811
7. Mulliken RS, Person WB (1962) Ann Rev Phys Chem 13:107
8. Weiss J (1942) J Chem Soc :245
9. Brackmann W (1949) Rev Trav Chim 68:147
10. Dewar MJS (1945) Nature 156:784; Dewar MJS (1946) J Chem Soc 777; Dewar MJS (1949) The electronic theory of organic chemistry. Oxford Clarendon Press, London
11. Haaland A (1989) Angew Chem 101:1017; Haaland A (1989) Angew Chem Intern Ed Engl 28:992
12. Dewar MJS, Lepley AR (1961) J Am Chem Soc 83:4560
13. Kopp RW, Bond AC, Parry RW (1976) Inorg Chem 15:3042
14. Steglich W, Höfle G (1969) Angew Chem 81:1001; Steglich W, Höfle G (1969) Angew Chem Intern Ed Engl 8:981
15. Weiß R, Engel S (1992) Angew Chem 104:239; Weiß R, Engel S (1992) Angew Chem Intern Ed Engl 31: 216
16. Schoeller WW, Schneider R (1997) Chem Ber 130:1013
17. Berlekamp UH, Jutzi P, Mix A, Neumann B, Stammler HG, Schoeller WW (1999) Angew Chem 111:2071; Berlekamp UH, Jutzi P, Mix A, Neumann B, Stammler HG, Schoeller WW (1999) Angew Chem Intern Ed 38:2048
18. Chuit C, Corriu RJP, Mehdi A, Reyé C (1993) Angew Chem 105:1372; Chuit C, Corriu RJP, Mehdi A, Reyé C (1993) Angew Chem Intern Ed Engl 32:1311
19. Belzner J, Schär D, Kneisel BO, Herbst-Irmer R (1995) Organometallics 14:1840
20. Lickiss P (1998) In: Rappoport Z, Apeloig Y (eds) The chemistry of organic silicon compounds, vol 2, part 1. Wiley, pp 557
21. Corey JY, West R (1963) J Am Chem Soc 85:4034; Hensen K, Zengerly T, Pickel P, Klebe G (1983) Angew Chem 95:739; Hensen K, Zengerly T, Pickel P, Klebe G (1983) Angew Chem Intern Ed Engl 22:725; Bassindale AR, Stout T (1984) J Chem Soc Chem Comm 1387; Bassindale AR, Stout T (1985) Tetrahedron Lett 26:3403; Chojnowski J, Stanczyk W (1990) Adv Organomet Chem 30:243; Kira M, Hino T, Sakurai H (1993) Chem Lett 153; Bahr SR, Boudjouk P (1993) J Am Chem Soc 115:4514
22. Aylett BJ, Campbell JM (1967) J Chem Soc Chem Commun 159
23. Lambert JB, Zhang S, Stern CL, Huffman JC (1993) Science 260:1917
24. Jutzi P, Bunte EA (1992) Angew Chem 104:1636; Jutzi P, Bunte EA (1992) Angew Chem Intern Ed Engl 31:1605
25. Steinberger HU, Müller T, Auner N, Maerker C, von R. Schleyer P (1997) Angew Chem 109:667; Steinberger HU, Müller T, Auner N, Maerker C, von R. Schleyer P (1997) Angew Chem Intern Ed Engl 36:626
26. Maerker C, von R. Schleyer P (1998) In: Rappoport Z, Apeloig Y (eds) The chemistry of organic silicon compounds, vol 2, part 1. Wiley, pp 513
27. Ottoson CH, Cremer D (1996) Organometallics 15:5309
28. Musher JI (1969) Angew Chem 81:68; Musher JI (1969) Angew Chem Intern Ed Engl 8:54
29. Schoeller WW, Dabisch T (1987) Inorg Chem 26:1081
30. Schoeller WW, Dabisch T, Busch T (1987) Inorg Chem 26:4383
31. Schoeller WW (1990) In: Regitz M, Scherer OJ (eds) Multiple bonds and low coordination in phosphorus chemistry. Georg Thieme, Stuttgart

32. Fluck E (1980) Top Phosphorus Chem 10:193
33. Cowley AH, Kemp RA (1985) Chem Rev 85:367
34. Sanchez M, Mazieres MR, Lamandé, Wolf R (1990) In: Regitz M, Scherer O (eds) Multiple bonds and low coordination in phosphorus chemistry. Georg Thieme, Stuttgart, D1, p 129
35. Gudat D (1997) Coord Chem Rev 163:71
36. Schoeller WW, Tubbesing U (1995) J Mol Struct (Theochem) 343:49
37. Gudat D (1998) Eur J Inorg Chem 1087
38. Niecke E, Kröher R (1976) Angew Chem 88:758; Niecke E, Kröher R (1976) Angew Chem Intern Ed Engl 15:692
39. Niecke E, Kröher R (1979) Z Naturforsch B 34:837
40. Niecke E, Nieger M, Reichert F (1988) Angew Chem 100:1781; Niecke E, Nieger M, Reichert F (1988) Angew Chem Intern Ed Engl 27:1715
41. Buenker RJ, Bruna PJ, Peyerimhoff SD (1980) Isr J Chem 19:309
42. Curtis RD, Schriver MJ, Wasylishen RE (1991) J Am Chem Soc 113:1493
43. Schoeller WW, Busch T, Niecke E (1990) Chem Ber 123:1653
44. Chernega AN, Antipin MY, Struchkov YT, Ruban AV, Romanenko V (1989) Zh Obshch Khim 30:105
45. Niecke E, Detsch R, Nieger M, Reichert F, Schoeller WW (1993) Bull Soc Chim Fr 130:25
46. Drapailo AB, Chernega AN, Romanenko VD, Madhouni R, Sotiropoulos J-M, Lamandé L, Sanchez M (1994) J Chem Soc Dalton Trans 2925
47. Niecke E, David G, Detsch R, Kramer B, Nieger M, Wenderoth P (1993) Phophorus Sulfur Silicon 76:25
48. Gudat D, Schiffner HM, Nieger M, Stalke D, Blake AJ, Grondey H, Niecke E (1992) J Am Chem Soc 114:8857
49. Burford N, Clyburde JAC, Bakshi PK, Cameron TS (1995) Organometallics 14:1578
50. David G, Niecke E, Nieger M (1992) Tetrahedron Lett 33:2335
51. David G, Niecke E, Nieger M, von der Gönna V, Schoeller WW (1993) Chem Ber 126:1513
52. Curtis JM, Burford N, Parks TM (1994) Org Mass Spectrom 29:414
53. Gudat D, Hoffbauer W, Niecke E, Schoeller WW, Fleischer U, Kutzelnigg (1994) J Am Chem Soc 116:7325
54. Burford N, Cameron TS, Clyburne JAC, Eichele K, Robertson KN, Sereda S, Wasylishen RE, Whitla WA (1996) Inorg Chem 35:5460
55. Niecke E, Nieger M, Reichert F, Schoeller WW (1988) Angew Chem 100:1779; Niecke E, Nieger M, Reichert F, Schoeller WW (1988) Angew Chem Intern Ed Engl 27:1713
56. Burford N, Cameron TS, Robertson KN, Phillips AD, Jenkins HA (2000) J Chem Soc Chem Comm 2087
57. Blättner M, Ruban AV, Gudat D, Nieger M, Niecke E (1999) Phosphorus Sulfur Silicon 147:1578
58. Blättner M (1998) Doctoral dissertation, Bonn, Germany
59. Blättner M, Nieger M, Niecke E, Schoeller WW (unpublished)
60. Glaser R (1989) J Phys Chem 93:7993
61. Glaser R, Horan CJ, Choy GS-C, Harris BL (1992) J Phys Chem 96:3689
62. Glaser R, Horan CJ, Choy GS-C, Harris BL (1993) Phosphorus Sulfur Silicon 77:73
63. Glaser R, Horan CJ (1995) J Org Chem 60:7518
64. Nguyen MT, Fitzpatrick NJ (1988) Chem Phys Lett 146:524
65. Busch T, Schoeller WW (1992) Chem Phys Lett 200:26
66. Pfister-Guillouzo G, Chrostowska A, Sotiropoulos J-M, Romanenko VD (1998) Eur J Inorg Chem:1821
67. Niecke E, Streubel R, Nieger M, Stalke D (1989) Angew Chem 101:1708; Niecke E, Streubel R, Nieger M, Stalke D (1989) Angew Chem Intern Ed Engl 28:1673
68. Romanenko VD, Rudzevich VL, Rusanov EB, Chernega AN, Senio A, Sotiropoulos J-M, Pfister-Guillouzo G, Sanchez M (1995) J Chem Soc Chem Comm 1383
69. Busch T, Schoeller WW, Niecke E, Nieger M, Westermann H (1989) Inorg Chem 28:4334
70. David G, Niecke E, Nieger M, Radseck J (1994) J Am Chem Soc 116:2191
71. Laali KK, Wesdemiotis C, Polce M, Beranová S (1995) J Chem Soc Perkin 2 1191

72. See Regitz M (1991) Phosphaalkynes. In: Regitz M, Scherer OJ (eds) Multiple bonds and low coordination in phosphorus chemistry. Thieme, p 58
73. Niecke E, Radzeck J, Schoeller WW (unpublished results)
74. Clyburne JAC, Schriver MJ (1996) Inorg Chem 35:3062
75. Laali KK, Geissler B, Regitz M, Houser JJ (1995) J Org Chem 60:6362
76. Schoeller WW, Strutwolf J (1994) J Mol Struct (Theochem) 305:127
77. Laali KK, Geissler B, Wagner O, Hoffmann J, Armbrust R, Eisfeld W, Regitz M (1994) J Am Chem Soc 116:9407
78. Avent AG, Cloke FGN, Flower KR, Hitchcock PB, Nixon JF, Vickers DM (1994) Angew Chem 106:2406; Avent AG, Cloke FGN, Flower KR, Hitchcock PB, Nixon JF, Vickers DM (1994) Angew Chem Intern Ed Engl 33:2330
79. Nixon JF (1995) Coord Chem Rev 145:201
80. Nguyen MT, Vansweevelt H, Vanquickenborne LG (1992) Chem Ber 125:923
81. Eisfeld E, Regitz M (1998) J Org Chem 63:2814
82. Smith D, McIntosh BJ, Adams NG (1989) J Chem Phys 90:6213
83. Zimmerman JA, Bach SBH, Watson CH, Eyler JR (1991) J Phys Chem 95:98
84. Niecke E, Leuer M, Nieger M (1989) Chem Ber 122:453
85. Schoeller WW, Haug W, Strutwolf J (1993) Bull Soc Chim Fr 130:636
86. Gleiter R, Pfeifer K-H, Baudler M, Scholz G, Wettling T, Regitz M (1990) Chem Ber 123:757
87. Wettling T, Schneider J, Wagner O, Kreiter CG, Regitz M (1989) Angew Chem 101:1035; Wettling T, Schneider J, Wagner O, Kreiter CG, Regitz M (1989) Angew Chem Intern Ed Engl 28:1013
88. Laali KK, Geissler B, Regitz M, Houser JJ (1995) J Org Chem 60:47
89. Laali KK, Regitz M, Birkel M, Stang PJ, Crittell CM (1993) J Org Chem 58:4105
90. Xu CS, de Beer E, Neumark DM (1996) J Chem Phys 104:2749
91. Cai Z-L, Hirsch G, Buenker RJ (1996) Chem Phys Lett 255:350
92. Schoeller WW, Haug W, Strutwolf J, Busch T (1996) J Chem Soc Faraday Trans 2 92:1751
93. Weiss R, Engel R (1992) Angew Chem 104:239; Weiss R, Engel R (1992) Angew Chem Intern Ed Engl 31:216
94. Meisel M, Lönnecke P, Grimmer AR, Wulff-Molder D (1997) Angew Chem 109:1940; Meisel M, Lönnecke P, Grimmer AR, Wulff-Molder D (1997) Angew Chem Intern Ed Engl 36:1869
95. Blättner M, Nieger M, Ruban A, Schoeller WW, Niecke E (2000) Angew Chem 112:2876; Blättner M, Nieger M, Ruban A, Schoeller WW, Niecke E (2000) Angew Chem Intern Ed Engl 39:2768
96. Schmidpeter A, Jochem G, Klinger C, Robl C, Nöth H (1997) J Organomet Chem 529:87
97. Schoeller WW, Rozhenko AB (unpublished results)

Top Curr Chem (2003) 229: 95–119
DOI 10.1007/b11152

Phosphinidenes

Koop Lammertsma

Department of Chemistry OAC, Faculty of Sciences, Vrije Universiteit, de Boelelaan 1083, 1081 HV Amsterdam, The Netherlands. *E-mail: Lammert@chem.vu.nl*

Of the low-valent organophosphorus compounds phosphinidenes occupy a special place. In 'free' phosphinidenes the phosphorus atom carries a single substituent. Such species prefer a triplet electronic ground state. Little is known about these species experimentally in contrast to those that carry a transition metal group. Depending on the ancillary ligands these carbene-like reagents can have electrophilic or nucleophilic properties. The comparison with Fischer-type and Schrock-type carbene complexes is tempting. In this survey, an analysis is given on the electronic characteristics of the phosphinidenes with and without transition metal groups. Synthetic approaches to the terminal complexed phosphinidenes are summarized and their reactivities are evaluated.

Keywords. Phosphinidene, P chemistry, Transition metal, Reactive intermediate, Fischer, Schrock

1 Introduction

Phosphinidenes [1] are low-valent organophosphorus compounds that have attracted attention since the early 1980s when they were first discovered [2]. They are known in two classifications, one being the six-electron singly substituted phosphorus species (**A**) and the other in which the phosphorus atom carries an additional η^1-stabilizing group, typically, but not necessarily, a transition metal group (**B**). Much has been learned about the reactivities of the complexed phosphinidenes **B**, particularly those that have electrophilic properties. Concise reviews have highlighted the synthetic potential of the reactive electrophilic intermediates [3,4] and the structural properties of the more robust nucleophilic ones [5]. The flurry of activities in this area of low-valent organophosphorus compounds has been fueled by the recognition that they behave like the corresponding hydrocarbons. Phosphorus is even slightly more electropositive than carbon. Because of this diagonal relationship, phosphorus has also been coined as the carbon copy [6]. In this survey we elaborate on the distinction between different types of phosphinidenes and highlight their special properties.

A **B**

Phosphinidenes (R-P) differ from other low-coordinate organophosphorus compounds, such as phosphaalkynes (R-C≡P), phosphaalkenes ($R_2C{=}PR$), and phosphaaromatics, in that the phosphorus atom carries only a single σ-bonded substituent [7–9]. They relate to carbenes, nitrenes, and silylenes and likewise can exist as singlet and triplet species. The advances that led to stable carbenes [10, 11] and silylenes [12] stimulated an exploration of the chemistry of phosphinidenes.

Phosphinidenes differ from carbenes because of the additional lone pair. This lone pair enables interactions with, e.g., a transition metal group for increased stability, while maintaining carbene-like behavior. These terminal η^1-complexed phosphinidenes differ from the μ_2-, μ_3-, and μ_4-complexes, which are not part of this survey. Phosphinidenes that are stabilized by a transition metal group also relate to carbene complexes. A distinction in Fischer and Schrock-type complexes has been advanced to distinguish phosphinidene complexes with nucleophilic properties from those that are electrophilic [13]. In this survey we address this topic in more detail.

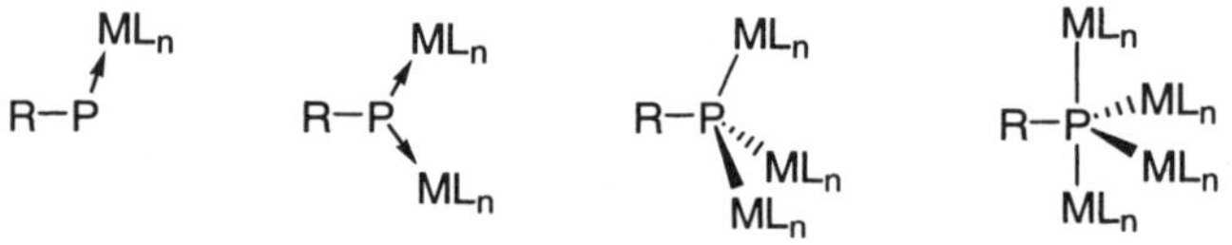

η^1 μ_2 μ_3 μ_4

2 'Free' Phosphinidenes

Phosphinidenes have either a singlet ground state with two lone pairs and an empty p-orbital on the phosphorus atom or a triplet ground state in which the phosphorus has instead one lone pair and two singly occupied p-orbitals. Not surprisingly, the electronic preference, i.e., the singlet-triplet energy gap (ΔE_{ST}) and thus the stability and reactivity of a phosphinidene, is determined by its substituent.

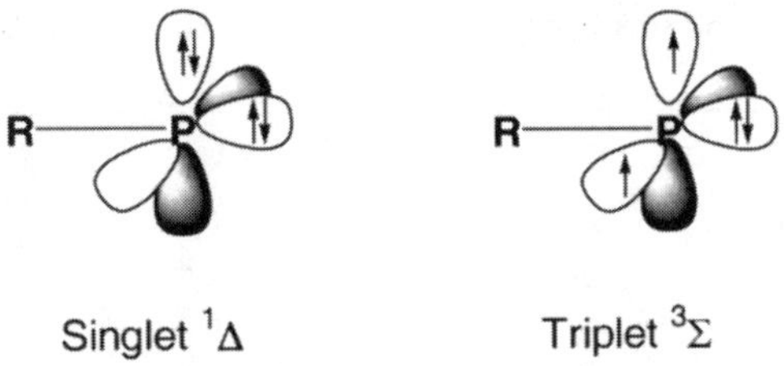

Phosphinidenes much prefer a triplet ground state, which is in striking contrast to silylenes for which a triplet only becomes viable on enforcing large angles between its substituents [14]. The experimentally determined ΔE_{ST} ($^3\Sigma^-$–$^1\Delta$) difference for PH amounts to 22 kcal/mol [15] with even larger theoretical estimates, ranging from 28 kcal/mol (QCISD(T)/6-311++G(3df,2p)) [16] to 41 kcal/mol (MP2/LanL2DZ) [13]. For comparison, SiH_2 prefers a singlet ground state by as much as 21 kcal/mol [17], while CH_2 and NH ($^3\Sigma^-$) have triplet ground states with ΔE_{ST} values of 9 [18] and 36 kcal/mol [19]. Simple alkyl or aryl substitution hardly influences ΔE_{ST}, except for the nitrene [20], and is still estimated at a formidable 24 kcal/mol for PhP [21, 22]. Gaspar and coworkers [23] obtained an EPR spectrum for MesP by irradiation of phosphirane **1** in a glassy matrix (77 K) and thereby confirmed its favored triplet state. Photolysis in the presence of acetylenes and olefins gave the three-membered ring structures phosphirenes and phosphiranes, respectively, which suggests MesP to be electrophilic [24]. Photolysis of phospholenes [23], phosphorus bis(azides) [25], and diphosphenes [26], and thermolysis of triazaphospholenes [27] and oxazaphospholenes [27] presumably also render phosphinidenes, but the synthetic applicability of these methods has remained very limited. Triplet phosphinidenes are considered less desirable synthons due to their anticipated lack of selectivity.

Stabilization of the singlet ground state of phosphinidenes can be induced by proper substitution. Such an approach has been applied to carbenes with great

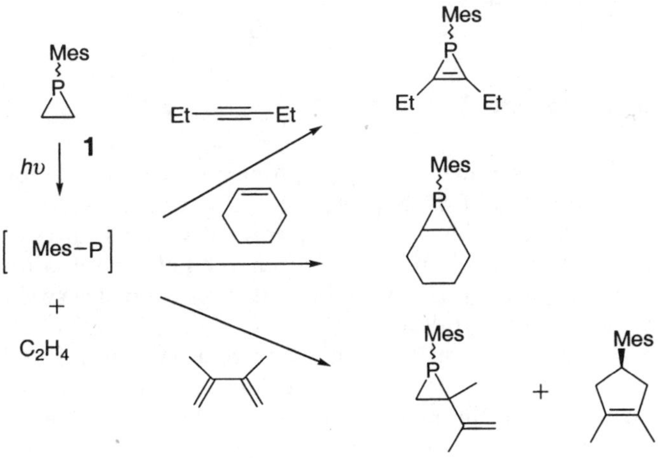

Table 1. Bond lengths (Å) and singlet-triplet energy differences (in kcal/mol) for triplet phosphinidenes

	P-H	P-Me	P-Ph	P-OH	$P-NH_2$	$P-PH_2$
d_{P-R} [a]	1.474	1.855 [b]	1.818	1.700	1.748	2.199 [b]
ΔE_{ST} (BP) [a]	–34.8	–30.1 [b]	–24.3	–19.7	–3.4	–6.9 [b]
ΔE_{ST} (QCI) [b]	–28.0	–26.0	–	–16.9	–1.2	–1.2

[a] [21].
[b] [16, 28].

success, as illustrated by the stable Arduengo- and Bertrand-type carbenes [10, 11]. Theoretical calculations show the same effect for phosphinidenes [16, 21, 28]. Thus, amino and phosphido substituents give a significant reduction in ΔE_{ST} values (Table 1), suggesting that the singlet species may be experimentally accessible. Electropositive groups have the largest effect because they enable a more favorable distribution of electron density due to their π-interaction with the phosphinidene center. Such effects are well known to stabilize carbenes [29]. However, we are unaware of experimental work in the condensed phase on free singlet phosphinidenes.

Theoretical studies have been devoted to the reaction of the parent singlet phosphinidene ^1PH with ethylene, water, hydrogen sulfide, hydrogen fluoride, and hydrogen chloride. The reaction coordinate for the highly exothermic formation of phosphirane from ^1PH and C_2H_4 (75 kcal/mol at CIPSI/4–31G*) showed no enthalpic barrier [30]. However, the exothermic insertion of ^1PH into the X-H bond of OH_2, SH_2, HF, and HCl occurs with small negative activation energies [31]. The initial interaction with these solvent molecules can be described as donor-acceptor complexes, which have very small barriers for hydrogen migration to the thermodynamically much more stable phosphines. Only the $HP-SH_2$ complex shows some ylide character. These reaction profiles for ^1PH are very similar to those for $^1CH_2$ [32], $^1SiH_2$ [33], and ^1NH [34], which also have negative activation energies for X-H bond insertion.

3
Phospha-Wittig Reagents

Phosphinidenes can be stabilized by a phosphine group. Such compounds relate to the Staudinger complexes $R_3P=NR'$ and the Wittig reagents $R_3P=CR'_2$. The two resonance structures of the phospha-Wittig reagent are illustrative of its multiple bonded character between the differently charged phosphorus atoms. A theoretical study on the parent P_2H_4 showed that this results in a very short PP distance of only 2.087 Å for the ylide at MP2/6–31G* [35]. This ylide is 27.6 kcal/mol less stable than $H_2P=PH_2$ at MP4. Stabilization of a phosphinidene by an ammonium group is similarly effective as in the Staudinger complex where the nitrene is stabilized by a phosphine group, but the NP bond distance of $HP-NH_3$ is much longer. An indication of the potential for generating phosphinidenes from these ylid structures can be extracted from an early study on phosphanylphosphinidene **2** [36]. This species, thermally generated from phosphanylidene-σ^4-phosphorane **3**, reacts with olefins to give phosphiranes, thereby illustrating its electrophilic carbene-like nature.

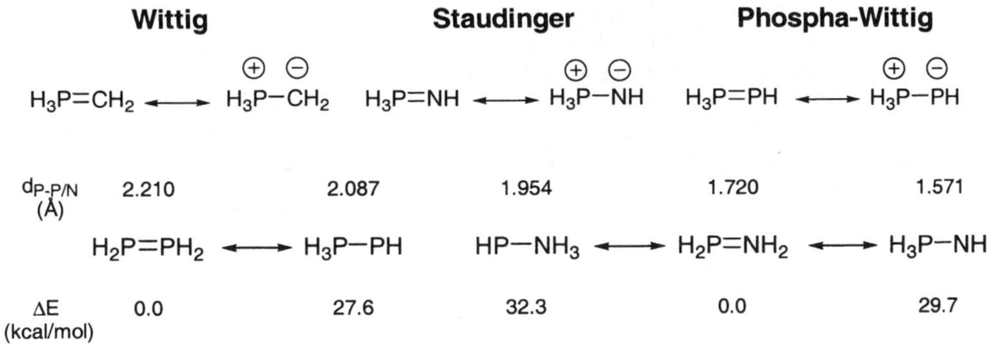

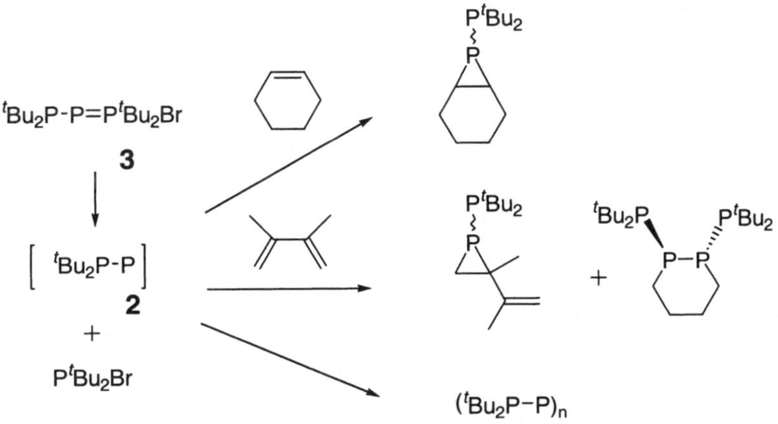

Recent studies have shown that phospha-Wittig reagents can behave like common Wittig reagents, besides their potential dissociation that renders a free phosphinidene. Access to the reagents from simple precursors has recently been expanded by a simple Zn-mediated condensation of dihalophosphines with phosphines as shown in [37]. The earlier Mg-mediated condensation of Mes*PCl_2 has been extensively used to generate diphosphenes [38]. Their formation can be explained by assuming the intermediacy of a free phosphinidene (Mes*P), but a recent study questions this hypothesis [39]. Decomposition of ArP=PR_3 **4**, which is promoted by oxygen, leads to the formation of diphosphenes ArP=PAr, presumably via an intermediate phosphinidene oxide ArPO [40]. Reaction of ArP=PMe_3 with aldehydes gives the corresponding phosphaalkenes, but no reactions have been observed with ketones [37]. The nucleophilic behavior of the phosphanylidene atom was confirmed in reactions with BF_3, alkyl and silyl iodides and tosylates, and acids [41]. A limitation of most phospha-Wittig reagents is their thermal instability that leads to dissociation into their constituents.

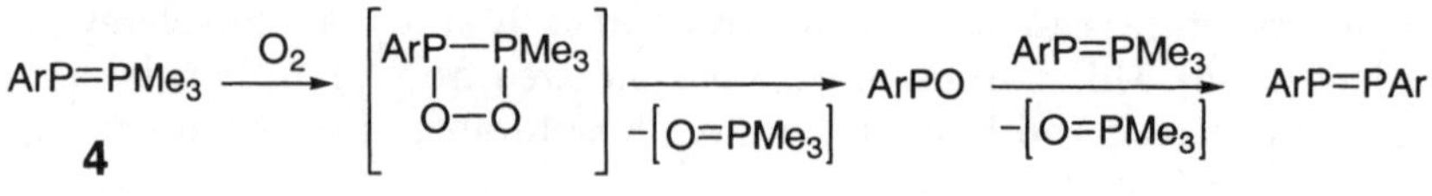

Phosphanylidenephosphoranes have been reviewed by Schmidpeter [42], while Shah and Protasiewicz [40] addressed their Wittig reactivity. Very recently, it was reported that photolysis of DmpP=PMe$_3$ (Dmp=2,6-Mes$_2$C$_6$H$_3$) gives DmpP=PDmp, while both Mes*P=PMe$_3$ and DipP=PMe$_3$ (Dip=2,6-Is$_2$C$_6$H$_3$) give intramolecular C-H and C-C insertions, respectively, which suggests the formation of free phosphinidenes [43]. Instead, a dimer is obtained from DmpP=PMes* [44].

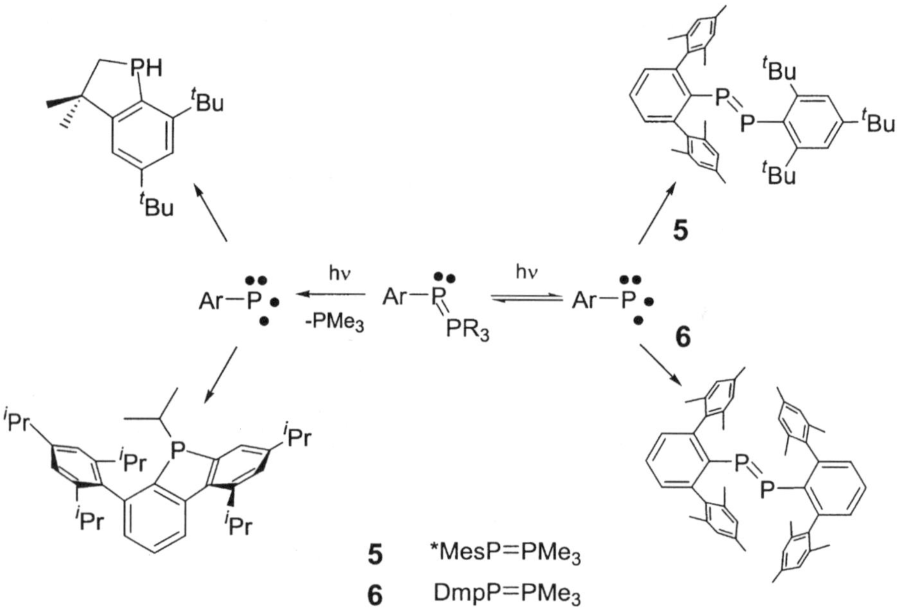

Potentially, phosphaalkenes can be precursors to phosphinidenes in the same manner that carbenes can be formed from alkenes. This latter metathesis route deserves more attention in light of the recognition that stable carbenes can be in equilibrium with their dimers [11]. However, a discussion on phosphaalkenes is outside the scope of the present survey.

4 Transition Metal Complexation

Adding a transition metal group to the phosphinidene in a η^1-fashion gives singlet species that are more palatable for use in chemical syntheses. Theoretical studies substantiated their preferred singlet nature, the extent of which depends on the substituent on phosphorus. When the transition metal group is M(CO)$_n$

the phosphinidenes are bent with the triplet species having larger angles between the substituent and the metal group than the singlet ones. For $(OC)_5W{=}PH$ this difference amounts to 20.8° and for the Cr complex this is 11.4°, both calculated with the economical BP density functionals with a large basis set [21]. Substitution of the singlet tungsten complex enlarges the angle from 102.5° for P-H to 121.7° for P-Ph. The ΔE_{ST} value for $(OC)_5Cr{=}PH$ amounts to 8.4 kcal/mol at the formidable CASPT2(12/12)/CASSCF [28] and to 9.5 kcal/mol using BP [21]; B3LYP with a smaller basis set gives a ΔE_{ST} of 6.0 kcal/mol. Electron donating substituents increase ΔE_{ST} (BP) for the $(OC)_5W{=}PR$ complex from 9.3 (R=H) to 22.7 kcal/mol when R is an alcohol group and to 29.9 kcal/mol for the amine derivative, thereby following the same trend found for the uncomplexed phosphinidenes (Table 2). Clearly, any unencumbered $(OC)_5M{=}PR$ complex will react as a singlet species.

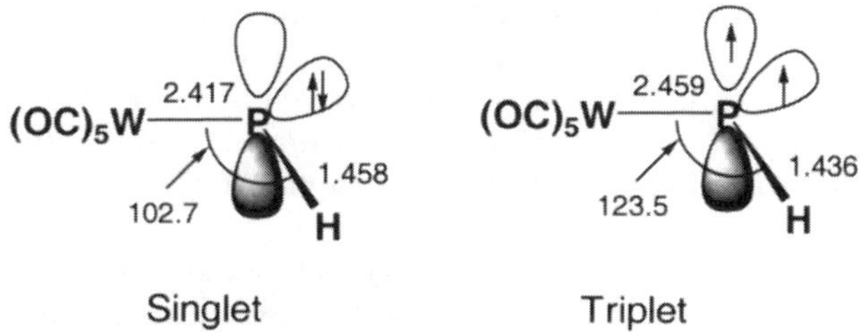

The interaction between the transition metal group and the phosphinidene is influenced by metal-phosphorus π-back-donation. Simply determining bond dissociation energies (BDE) for, for example, singlet $(OC)_5M{=}PR$ (M=Cr, Mo, W) is not adequate (Eq. 1 in Scheme 1), because the phosphinidene and $M(CO)_5$ fragments favor different electronic ground states. Hence, ΔE_{ST} for the presumably fast singlet-triplet relaxation of PR has to be added to the BDE_S of the singlet complex (boxed in Eq. 2 in Scheme 1), identified as BDE_{ST} in Table 2 [21]. Alternatively, the singlet-triplet promotion energy for singlet $(OC)_5M{=}PR$ can be added to the BDE_T of the triplet species (boxed in Eq. 3 in Scheme 1) and these are listed as BDE_{TS}. The two values differ slightly due to non-linear additivities.

Table 2. Singlet-triplet energy differences (ΔE_{ST}) and singlet (BDE_S), triplet (BDE_T), and composite ($BDE_{ST/TS}$) bond dissociation energies for phosphinidenes and their $W(CO)_5$ complexes, calculated with BP/TZP

Complex	ΔE or BDE	P-H	P-Ph	P-OH	P-NH_2
None	ΔE_{ST}	–34.8	–24.3	–19.7	–3.4
$W(CO)_5$	ΔE_{ST}	9.3	10.3	22.7	29.9
	BDE_S	77.8	67.4	71.8	62.4
	BDE_T	35.6	35.4	31.4	35.5
	BDE_{ST}	43.0	43.1	52.1	59.0
	BDE_{TS}	44.9	45.7	54.1	65.4

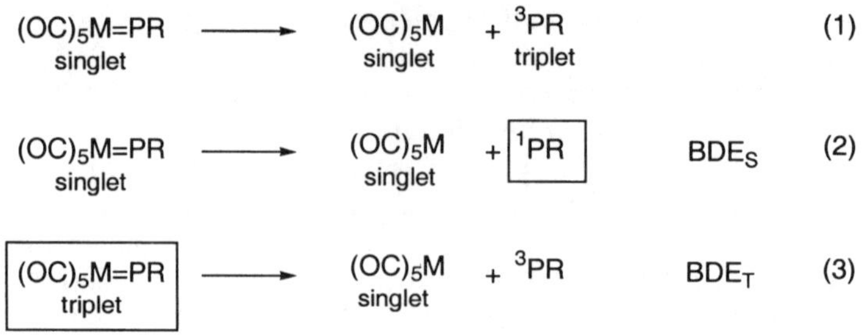

Scheme 1

These composite BDE values increase strongly on substituting the phosphorus atom with π-electron donating groups. However, this substitution effect does not relate to strengthening of the M=P bond, but rather reflects a reduction in the ΔE_{ST} of the free phosphinidene or conversely an increase in the ΔE_{ST} for the complexed phosphinidene. Still, as a point of reference, it is noted that similar CASPT2 and BP values of 41.1 and 40.5 kcal/mol, respectively, have been deduced for the dissociation of the parent Cr complex $(OC)_5Cr{=}PH$ into $Cr(CO)_5$ and ^3PH.

Multiple M=P bonding in $(OC)_5M{=}PR$ becomes evident with ADF's bond energy analysis in terms of electrostatic interactions, Pauli repulsion, and orbital interactions from which the σ,π-separation is obtained using a symmetry decomposition scheme [21]. For singlet $(OC)_5Cr{=}PR$, which has a BDE_{ST} of 40.5 kcal/mol, the σ- and π-components are 62.4 and 40.9 kcal/mol, respectively.

The amine substituent in the corresponding tungsten complex reduces ΔE_π to 21.3 kcal/mol, while the effect on ΔE_σ is negligible.

Depending on the ancillary ligands of the transition metal, the properties of terminal complexed phosphinidenes range from electrophilic to nucleophilic, in analogy with the well-known Fischer and Schrock type carbene complexes. Illustrative are the tungsten complexes $Cp_2W{=}PMes^*$, which is very stable and nucleophilic, and $(OC)_5W{=}PR$, which is only known as a transient and highly reactive electrophilic reagent.

A DFT study provided insight into the factors governing the philicity of the phosphinidene complex $ML_n{=}PH$ by varying the metal (M=Ti, Zr, Hf, V, Nb, Ta, Cr, Mo, W, Fe, Ru, Os, Co, Rh, Ir) and the ligands (L=CO, PH_3, Cp). The philicity is influenced by the spectator ligands L and hardly by the metal M [45]. Ligands with strong σ-donor capabilities on the metal increase the electron density on the phosphorus atom, raise the π*-orbital energy, and enhance its nucleophilicity. Conversely, strong π-acceptor ligands lower the charge on phosphorus, stabilize the π*-orbital energy, and enhance its electrophilicity. The extent of σ- and π-character of the $ML_n{=}PH$ bond, determined with the noted bond energy analysis and symmetry decomposition scheme for the orbital interactions, show the phosphinidenes to be strong π-acceptors and even stronger σ-donors. The interaction between ML_n and the phosphinidene is stronger for the second- and third-row transition metals than for those of the first-row for which the important interactions are shown in Table 3.

Table 3. Bond lengths (Å), bond dissociation energies (kcal/mol), σ- and π-bond strengths (kcal/mol), charges on phosphorus (e), and orbital energies (eV) for first row transition metal complexes ML_n=PH

ML_n=PH	Cp_2Ti	$Cp(PH_3)_3V$	$Cp(PH_3)Co$	Cp(CO)Co	$Cp(CO)_3V$	$(CO)_4Fe$
M=P	2.352	2.326	2.085	2.090	2.307	2.192
BDE	60.0	49.5	66.1	65.8	52.8	61.7
ΔE_σ	−91.0	−98.0	−88.0	−85.4	−94.3	−90.7
ΔE_π	−33.9	−34.2	−27.9	−28.6	−37.2	−32.2
Charge P	0.27	0.21	0.17	0.12	0.08	0.06
E(HOMO-ML_n)	−3.21	−3.47	−3.47	−4.39	−5.34	−5.60
E(LUMO)	−2.16	−2.93	−3.12	−3.72	−4.09	−4.41

Electrophilic and nucleophilic phosphinidene complexes have been related to the corresponding carbene complexes of which the Fischer-type is usually considered as a singlet-singlet combination and the Schrock-type as a triplet-triplet combination. However, both the strongly preferred triplet state of R-P and the M=P bond analysis suggest this schematic interpretation to be less appropriate for transition metal complexed phosphinidenes.

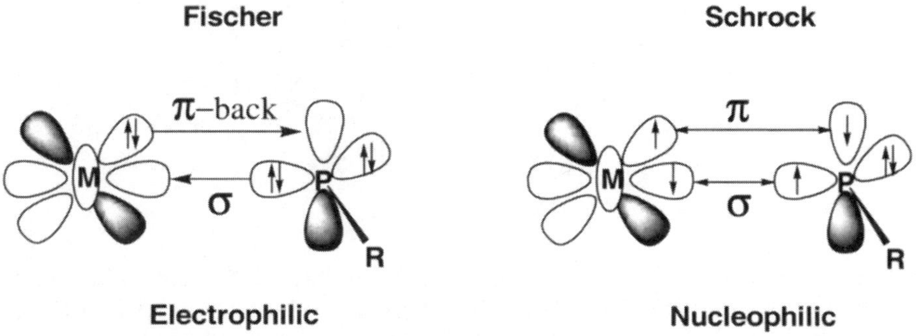

5 Electrophilic Phosphinidene Complexes

The most widely employed electrophilic phosphinidene complexes are generated from $M(CO)_5$ (M=W, Mo, Cr) complexed 7-phosphanorbornadienes **7**, which were reported by Marinetti et al. in 1982 [46]. They trapped the transient phosphinidene complex **8** at 110 °C (or at 55 °C if CuCl was used as catalyst) with olefins and alkynes to obtain three-membered phosphiranes and phosphirenes, respectively.

Cheletropic elimination of the in situ generated phosphinidene complex **8** from the 7-phosphanorbornadiene precursor is believed to be the rate-determining step before 1,2-cycloaddition occurs to the unsaturated hydrocarbon. Without catalysts these are first-order processes that depend only on the concentration of the precursor and not on that of any substrate [47]. The configu-

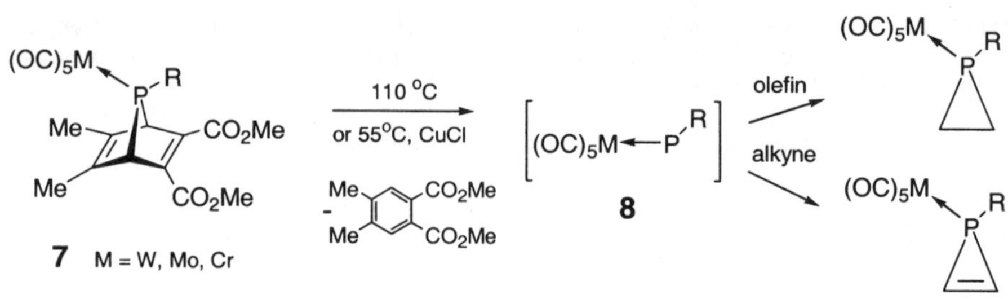

ration of Z- and E-substituted olefins is retained during the cycloaddition [48]. The small and negative Hammett reaction constants of –0.76 ($PhPW(CO)_5$) [49], –0.60 ($MePW(CO)_5$) [50], and –0.55 ($MeOPW(CO)_5$) [51] for CuCl-catalyzed addition to styrenes support the notion that the reactive intermediates are electrophilic carbene-like species. It is noted, however, that no intermediate $(OC)_5M{=}PR$ (M=W, Mo, Cr) phosphinidene complex has been observed spectroscopically. In the absence of trapping reagents, dimers, trimers, and tetramers are formed [47, 52, 108]. A recent computational study suggests that the electrophilic species in the CuCl activated decomposition is in fact an $(OC)_5W{=}PR\ldots CuCl$ complex [53]. Support for such a species can be inferred from dimer **9** that results from reaction with phospholenes [54].

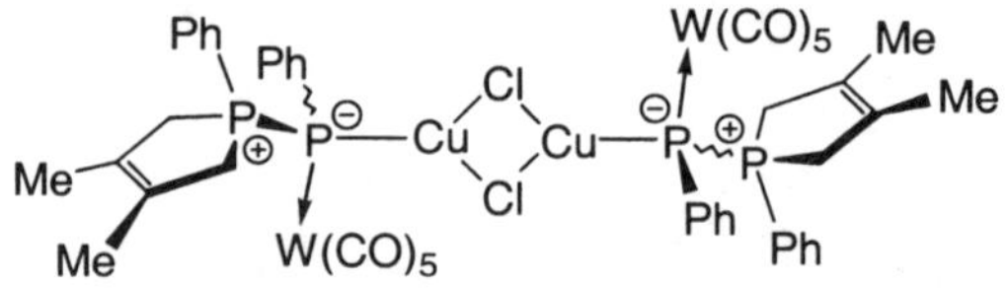

9

Amino-substituted phosphirane complexes undergo retro-cycloaddition at 70–90 °C to give the 'trappable' $(OC)_5WPNEt_2$ phosphinidene complex [55]. Other substituted phosphiranes are not suited to generate phosphinidenes because they require elevated temperatures, which typically induce thermal degradation instead.

Complexed azaphosphirenes have been used by the Streubel group to generate at 45–75 °C in situ phosphinidene complex $(OC)_5W{=}PCH(SiMe_3)_2$ (**10**) together with nitrilium ylid **11** [56].

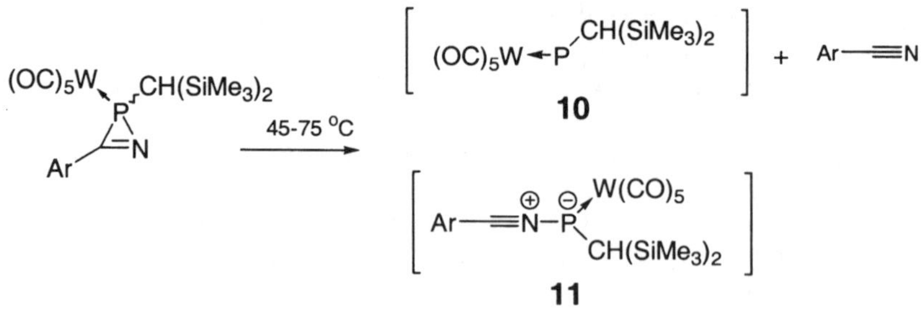

Cowley's group reported in 1987 the condensation of a dichlorophosphine with Collman's reagent to form stable complex **12**, for which they obtained an X-ray crystal structure [57]. However, the two nearly equally long P-N bond lengths of 1.777(7) and 1.764(7) indicate that **12** is not an unencumbered species.

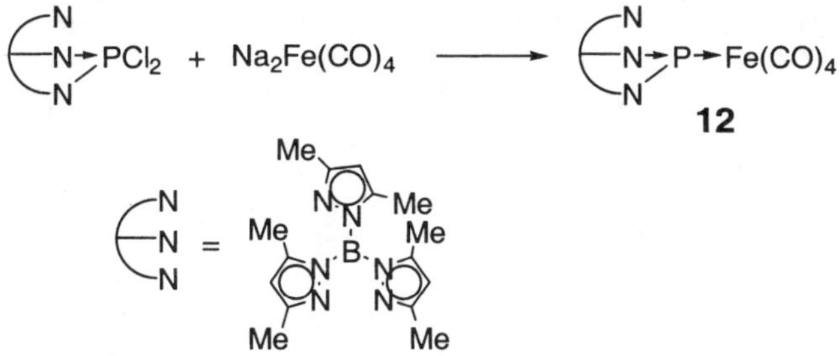

In 2001 Carty and coworkers reported the synthesis and crystal structures of several cationic phosphinidene complexes **13–15** having $AlCl_4^-$ as counterion [58–60]. These complexes all contain a single Cp* ligand and have M=P bond lengths that are longer than expected for double bonds due to the stabilizing effect of the N^iPr_2 group (Table 4). The extremely low field ^{31}P NMR chemical shifts of these phosphinidene complexes are characteristic. As of yet little information is available about their reactivity other than that the reaction of **14** with phenyl acetylene gives cationic metalaphosphacyclobutene complex **16** [42].

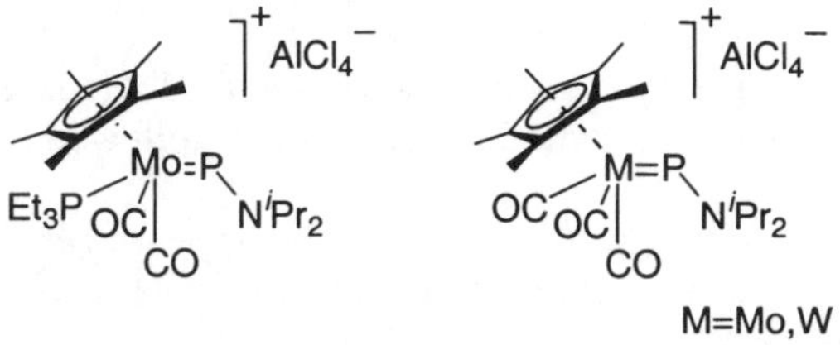

13 **14**

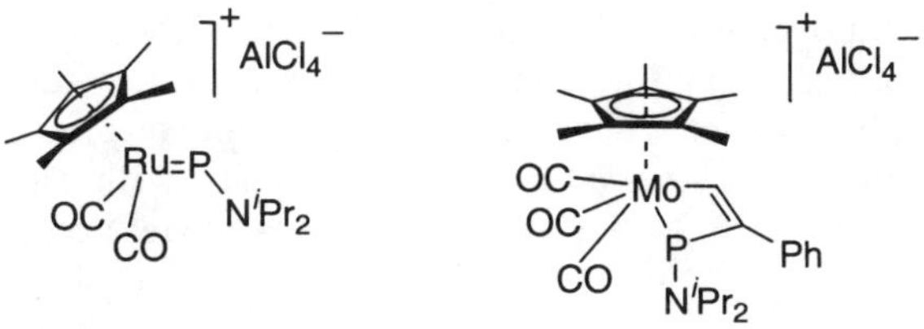

15 **16**

Recently, the group of Lammertsma developed an alternative route, using Collman's reagent and dichlorophosphine **17**, to generate $Fe(CO)_4$ complexed phosphinidene **18** in situ [61]. This reactive amino substituted species is trappable at about 0 °C with alkynes and terminal alkenes to give stable phosphirene

Table 4. M=P bond distances, M=P-R bond angles, and ^{31}P NMR chemical shifts of stable cationic terminal phosphinidene complexes

Cmpd	**13** (Mo)	**14** (Mo)	**14** (W)	**15** (Ru)
d(M=P)	2.3816(4)	2.4506(4)	2.4503(6)	2.2654(5)
α(M=P-R)	121.49(5)	118.79(5)	118.69(8)	119.58(6)
$\delta(^{31}P)$	957	1007.5	939.4	932

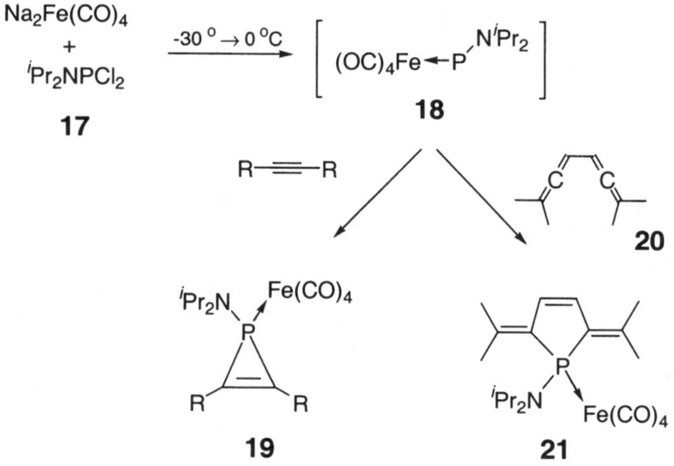

(**19**) and phosphirane complexes, respectively. Reaction with diallene **20** gives phospholene **21**, presumably via an intermediate 1,2-adduct followed by a 1,3-sigmatropic shift [62].

The first report on an iron complexed phosphinidene appeared in 1984 by Bertrand and coworkers [63]. They reported that abstraction of Cl^- from **22** with $Ph_3C^+PF_6^-$ at −90 °C presumably results in cationic phosphinidene complex **23**, which was based on its downfield ^{31}P NMR resonance at 954 ppm. Above −68 °C the phosphorus center inserts into a isopropyl C-H bond to give **24** as only isolable product. Dehalogenation with $AlCl_3$ resulted directly in the $AlCl_4^-$ salt of **24**.

A plethora of reactions has been reported for the in situ generated $(OC)_5W{=}PPh$, many of which have been the topic of previous reviews [2–4]. A typical account of the versatility of this reagent with olefinic and acetylenic bonds is visualized in the 'wagon wheel'. Only a brief account with selected ex-

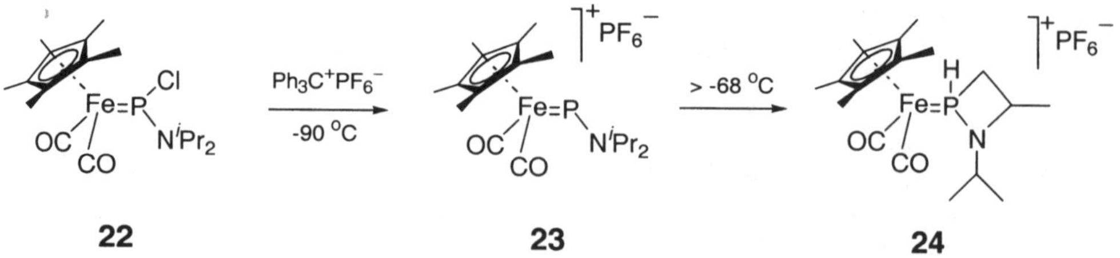

amples is given here. The best-known reaction is the olefin 1,2-cycloaddition that gives three-membered phosphiranes in which the configuration of the olefin is maintained [47]. Three-membered ring structures with saturated and unsaturated CCP, CNP [56, 64], COP [56], CPSi [65], CPS [66], CPP [67] have been synthesized either by direct addition of the phosphinidene complex to the C=X heteroolefinic bond or via an ylid intermediate. There is little or no selectivity for the addition to differently substituted olefins [68], which is in agreement with generating the phosphinidene complex being the rate determining step [47]. As noted, cycloaddition of ^{1}PH to ethylene occurs without barrier. Likewise, there is little selectivity in reactions with exocyclic double bonds [69]. These yield remarkably stable spiro compounds of which complexed phosphaspiropentane **25** is the smallest member [69]. Even a complexed bispirophosphirane, phospha[3]triangulene **26**, has been generated from the addition of biscyclopropylidene [70]. Addition to allenes results in complexed methylenephosphiranes **27** [71], which have been shown to undergo ring opening by both strong base [72]

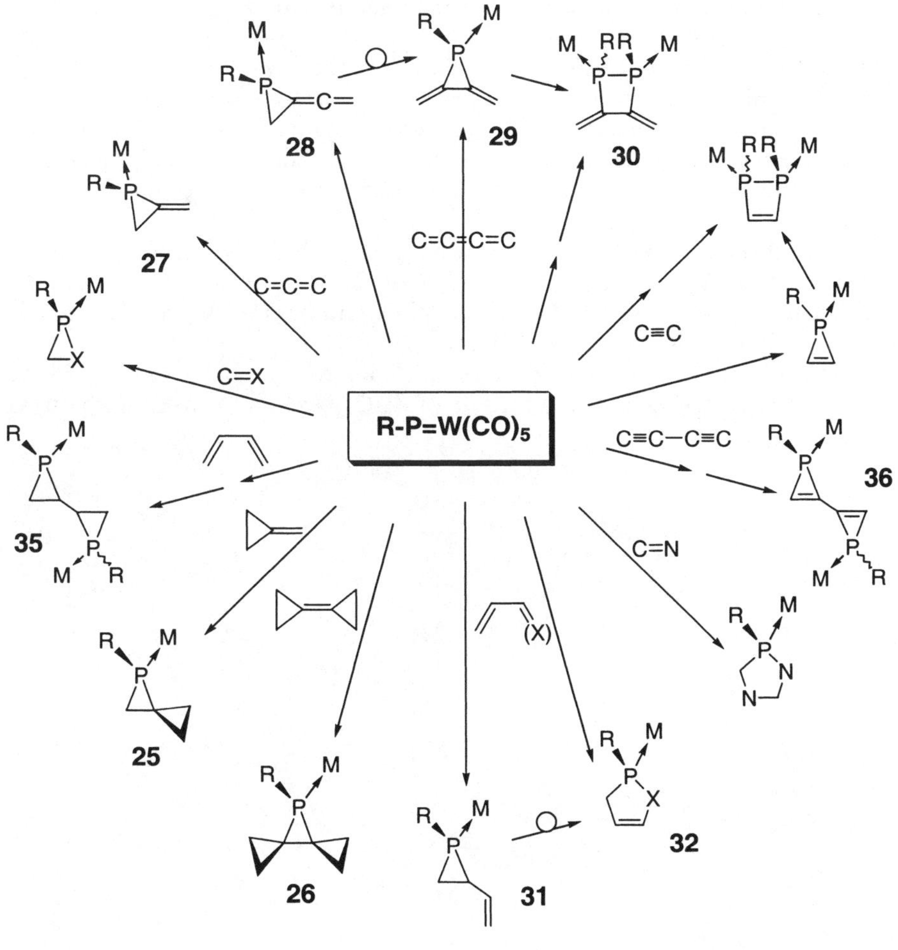

and $Ru_3(CO)_{12}$ [73]. Reaction with cumulenes occurs at its terminal double bonds giving complexed vinylidenephosphiranes **28**, which rearrange to the corresponding phospha[3]radialanes **29** [74]. Subsequently, the phosphinidene inserts in selected cases into the C-P bond to give **30**.

Phosphinidene $(OC)_5W{=}PPh$ adds to 1,3-dienes in both a 1,2- (major) and 1,4-fashion (minor), giving complexed vinylphosphiranes **31** and the thermodynamically favored phospholenes **32**, respectively [48]. The five-membered rings are the only products resulting from addition to α,β-unsaturated ketones [48] and 1-azadienes [75]. Phospholenes are also formed from the 1,2-adducts via a 1,3-sigmatropic shift. A study on 1-methoxy-1,3-cyclohexadiene suggests that this rearrangement occurs in a pericyclic manner, with inversion of the phosphorus configuration [76]. Instead, the 1,2-adducts of differently sized cyclic 1,3-dienes undergo only epimerization at the phosphorus center, which suggest a diradical process [77]. A kinetic study on the formation of 1,2- and 1,4-adducts **33** and **34** from a cisoid diene suggested a diradical pathway for the 1,3-shift of the 1,2-adduct and a concerted pathway for the CuCl catalyzed process [78]. There is a remarkable analogy between the rearrangements of complexed vinylphosphiranes with those of vinylcyclopropanes. This is supported by an extensive theoretical study, which concluded that neither the epimerization of $Cr(CO)_5$-complexed vinylphosphiranes **31** nor the rearrangement to phospholenes **32** occur via diradical intermediates [79]. Just as for the hydrocarbons, the transition states display diradical behavior and hence the rearrangements occur in a non-pericyclic manner, despite that the products seemingly result from a concerted process.

When 1,3-dienes remain in an anti orientation, inhibiting a [1,3]-shift of the first formed vinylphosphirane, a second phosphinidene addition can occur to give bisphosphiranes **35** [80]. Bis- and trisphosphirenes (**36–38**) result from addition to alkyne substituted phosphirenes (**39**) [81, 82].

So far, only one reaction with an aromatic substrate has been reported in the literature. This concerns the addition of $(OC)_5W{=}PPh$ in a 1,4-fashion to

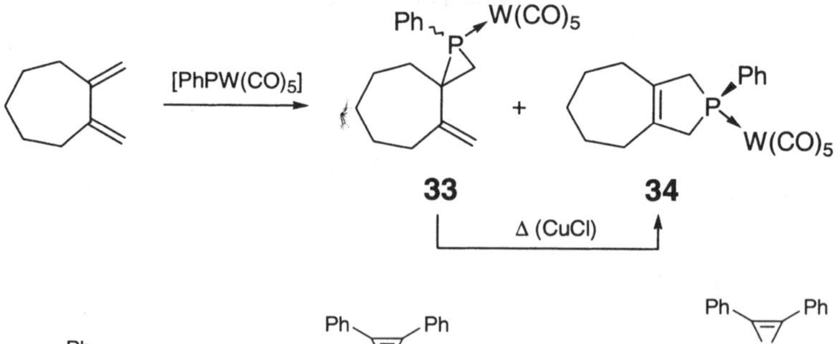

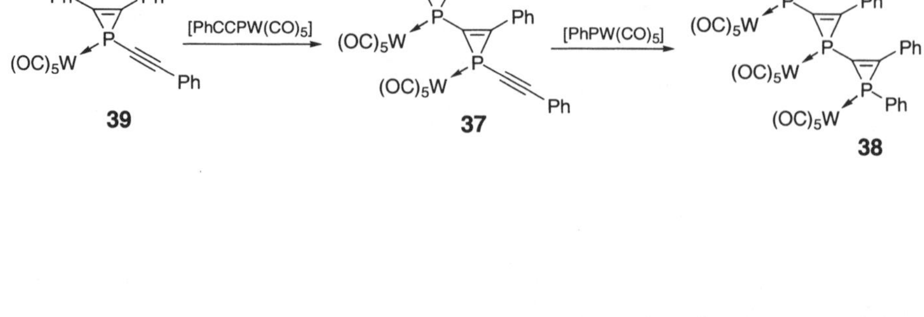

[5]metacyclophane [83]. In this remarkable reaction the phosphinidene complex is transferred from one benzene derivative to another with toluene as solvent. Removal of the transition metal group resulted in the as yet only known free 7λ^3-phosphanorbornadiene **40** [84]. The moderate stability of this system results from the unfavorable thermodynamics for retro-addition to the highly strained hydrocarbon.

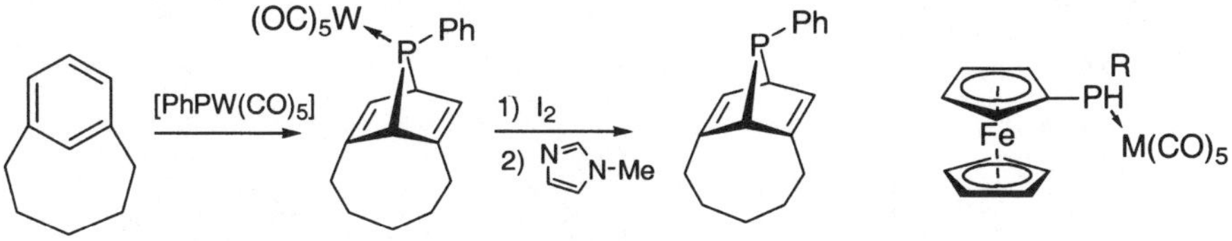

40 **41**

The only other reaction with an aromatic substance is the C-H insertion into ferrocene [85], giving **41**, which illustrates the highly electrophilic character of the phosphinidene complex. Other aromatic C-H insertions have been observed, but these likely occur by means of intermediate P,O- and P,N-ylids, such as the reaction of $(OC)_5W{=}PR$ with benzophenone and azobenzene that give **42** and **43**, respectively [56a, 86].

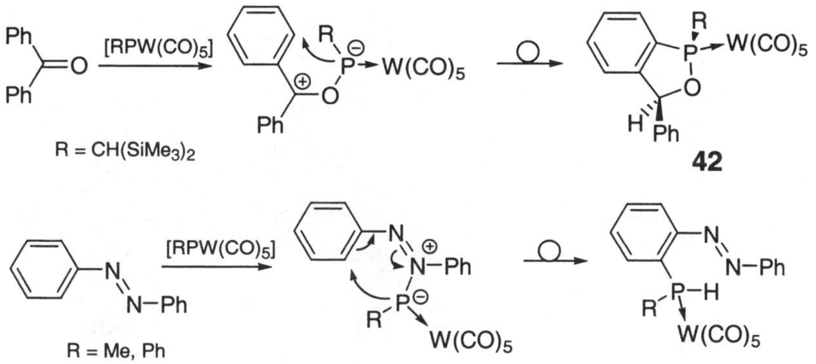

42

43

C-H insertion also occurs in the reactions with acetone and acetophenone, presumably through the rearrangement of transient OH-substituted phosphiranes [87]. C-C insertions occur for diketones to give **45** and have been postulated to occur via initial 1,2-addition to the conjugated enol **44** [87]. Diimines **46** also undergo C-C insertions [88]. Based on a theoretical evaluation, the products **47** are considered to result from a 2,3-sigmatropic rearrangement of initial formed P,N-ylids.

An intriguing [3,3]-phospha-Cope rearrangement has been suggested to underlie the formation of **49** [89]. The initial step is considered to be the addition of transient $(OC)_5W{=}P{-}C(CH_3){=}CH_2$ to 2,3-dimethyl-1,3-butadiene to give **48**,

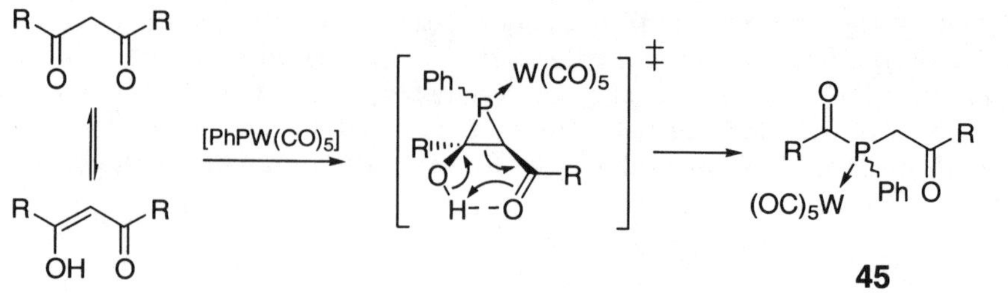

44 45

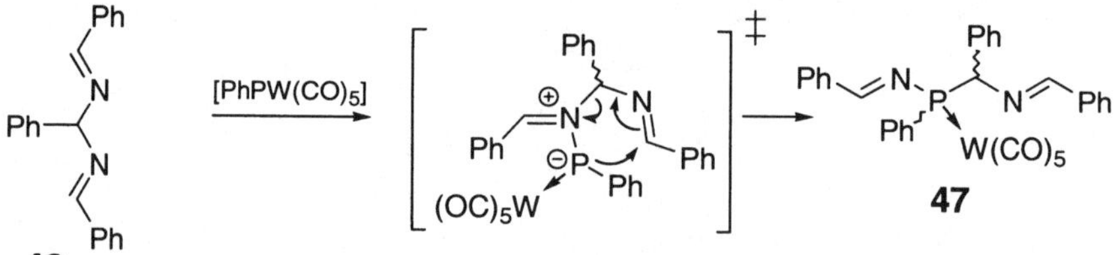

46 47

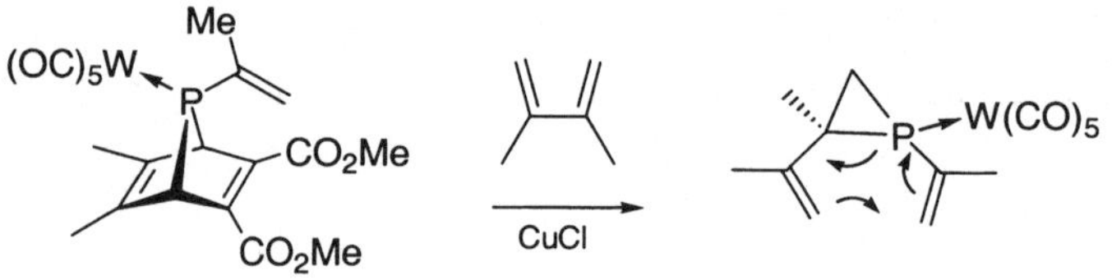

48

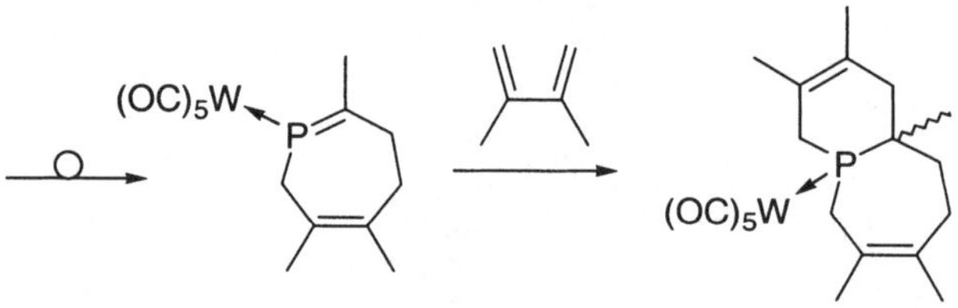

49

which then ring expands and subsequently undergoes a Diels-Alder addition with the diene.

Of the many reactions that have been performed with heteroolefinic systems we mention the addition to 1,3,5-triphosphabenzene, because of the special relationship between the 1,4-adduct **50** and the corresponding quadricyclane **51** [90].

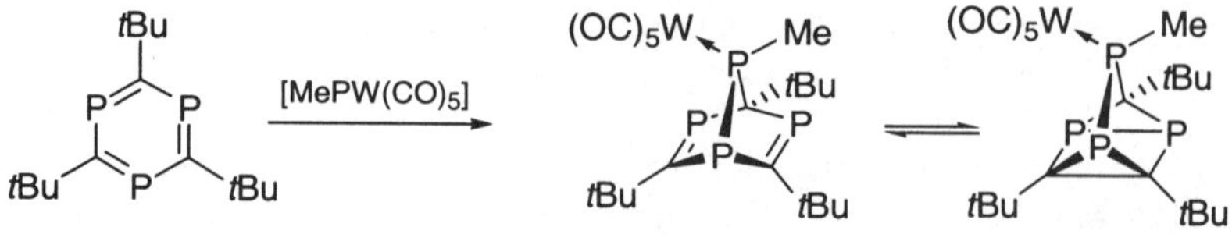

50 51

6
Nucleophilic Phosphinidene Complexes

Since the mid-1980s many more phosphinidene complexes have been prepared with nucleophilic than with electrophilic properties. An important contributing factor is their significant stability enabling even crystal structure determination. A 1997 review, focusing on structural, electronic, and spectroscopic aspects, suggested that the quest for terminal phosphinidenes is over [5]. Much has happened since, making it relevant to address some issues that are pertinent in the present context. The displayed structures are a simplified compilation of currently known complexes, all of which were characterized by crystal structures (Table 5). Common to all these structures is the presence of at least one π-donating ligand, which distinguishes these complexes from the electrophilic ones.

In 1987 Lappert et al. reported the first stable Mo and W complexed phosphinidenes **52** with two Cp ligands on the transition metal from the reaction of Mes*PCl with $Cp_2M(H)Li$ [91]. Niecke et al. approached the syntheses of similar compounds with instead a NHMes* substituent using Mes*NPCl and $Cp_2M(H)Li$

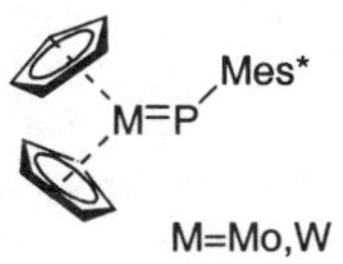

52

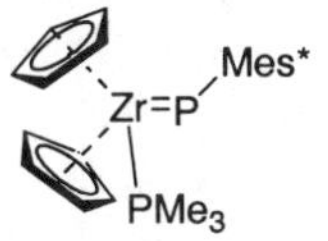

53

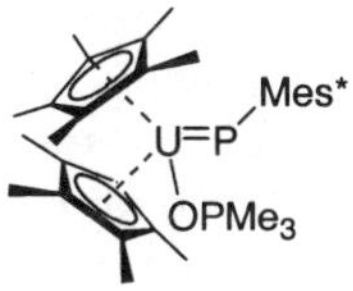

54

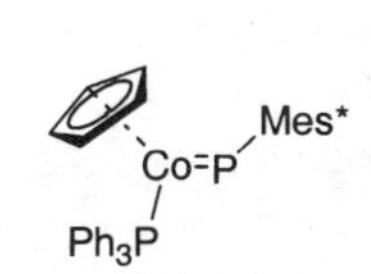

55

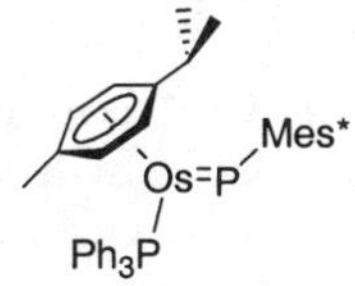

56

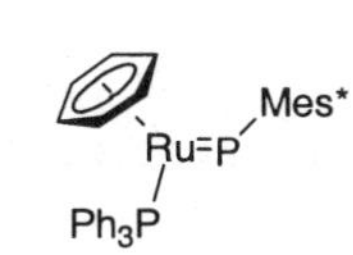

57

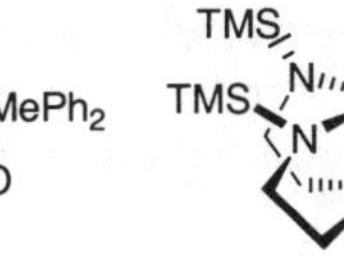

58

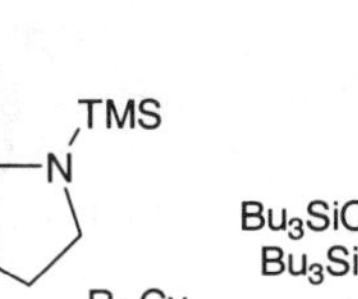

59

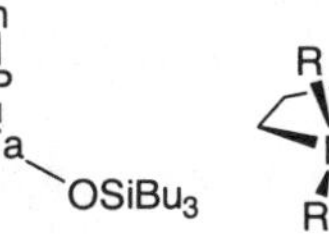

60

61

62

63

64

Table 5. M=P bond distances and ^{31}P NMR chemical shifts and J_{PP} coupling constants of stable nucleophilic terminal phosphinidene complexes

Cmpd	**52** (Mo)	**52** (W)	**53** (Zr)	**54** (U)	**55** (Ir)	**56** (Ir)	**57** (Rh)
d(M=P)	2.370(2)	2.349(5)	2.505(4)	2.562(3)	2.2121(5)	2.1783(8)	2.1903(4)
α(MPR)	115.8(2)	114.8(5)	101.4(1)	143.7(3)	113.73(7)	113.77(10)	113.94(5)
$\delta(^{31}P)$	799.5	661.1	792.4	71.1	686.6	804.6	867.6
$J_{PP/PM}$		–/153.5	23		102		59.4/68.6
Cmpd	**58** (Co)	**59** (Os)	**60** (Ru)	**61** (W)	**62** (Ta)	**63** (Ta)	**64** (Ni)
D(M=P)	2.1102(8)	2.2196(7)	2.1989(7)	2.169(1)	2.145(7)	2.317(4)	2.0772(9)
α(MPR)	109.00(9)	106.56(9)	107.80(7)	168.2(2)			
$\delta(^{31}P)$	866.9	667.5	845.9	193.0	209.8	334.6	970
$J_{PP/PM}$	76	80	40	50/649			134

as starting materials, but were only able to obtain ^{31}P NMR data (Mo: 663; W: 770) [92]. In the early 1990s an analogous Zr complex **53** with an additional stabilizing phosphine ligand was reported by the group of Stephan [93], which was followed by the synthesis of a U complex **54** that carries instead an additional $OPMe_3$ ligand [94]. The formation of these complexes have in common that LiCl (or KCl) is abstracted in the reaction of either the Li (or K) salt of the metal complex with a dichlorophosphine or conversely the halogenated metal complex with a Li (or K) phosphide. The dehalogenation approach is not always successful. For example, treatment of Cp^*_2HfX(PHR) with bases such as *n*BuLi, KH, and $NaN(SiMe_3)_2$ did not afford the desired phosphinidene complex [95] and likewise Me_3SiCl could not be eliminated from $Cp_2ZrCl\{P(SiMe_3)Mes^*\}$ [96]. It may well be that these attempts failed due to the absence of an additional stabilizing ligand, like PR_3 in **53** or OPR_3 in **54**. Without a phosphine ligand, Cp^*_2Zr=PMes* has only been detected by its ^{31}P NMR resonance at 537 ppm [93b], which is 254 ppm shielded from phosphinidene complex **53**. Majoral et al. reported similar ^{31}P NMR resonances at 465.2 and 438.2 ppm for respectively Cp_2Zr=PR and Cp^*_2Zr=PR (R=$C_6H_2(OMe)_3$), but could not exclude the formation of dimers [97].

The formation of the single Cp* containing phosphinidene complex $Cp^*(Cl)_2Ta$=PIs (Is=$C_6H_2(CHMe_2)_3$) by dehydrohalogenation of $Cp^*Cl_4Ta(PH_2Is)$ has been described by Hey-Hawkins and coworkers, including a crystal structure, but no details are yet available [98]. Treatment of the slightly less congested $Cp'(Cl)_2Ta$=PIs (Cp′=C_5H_4Me) with DBU results instead in a dimer, $[Cp'TaCl\{\mu\text{-}PIs\}]_2$ [99]. Similar dimers were obtained from the reaction of Cp^*TaCl_4 with LiPHR (R=Cy, *t*-Bu, Ph, Mes) [100]. Using a strong base induced H-migration enabled Malisch et al. to generate $Cp(CO)_2HW$=PMes* from $Cp(CO)_2W$=PHMes* [101]. Even though the compound was too sensitive to be isolated, its ^{31}P NMR chemical shift of 819.9 ppm (J_{PW}=123.1 Hz) is characteristic for a bent phosphinidene.

Using a dehydrohalogenation-ligation sequence, the group of Lammertsma generated phosphinidene complexes with group 8 (Ru, Os) and group 9 (Co, Rh,

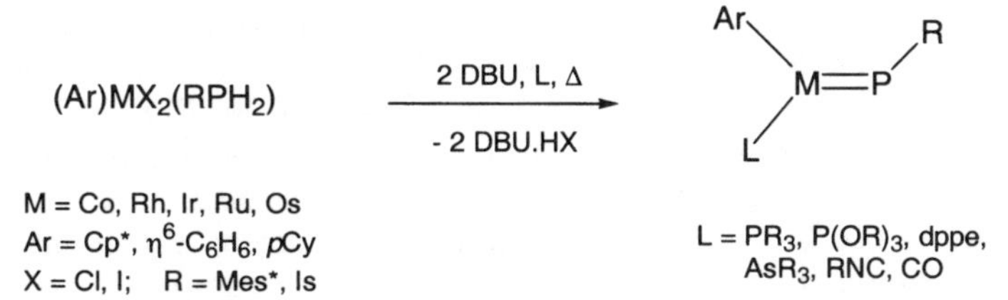

Ir) transition metals following the outlined reaction scheme [102–104]. Both E and Z conformers can be formed depending on the size of the substituent R (Mes* or Is) and the ligands L and Ar (Cp* for group 9; η^6-C_6H_6 or *p*Cy for group 8). For example, iridium complex **55** prefers the E form, while the less congested **56** has a Z configuration. Ir complexes Cp*(L)Ir=PR have been synthesized with L being PPh_3, PH_2Mes^*, PMe_3, $P(OMe)_3$, dppe, $AsPh_3$, *t*-BuNC, XyNC, and CO (Xy=2,6-dimethylphenyl) [102]. Those with the ligands PMe_3 and $P(OMe)_3$ are formed as E/Z-mixtures. The ^{31}P NMR chemical shifts and $^2J_{PP}$ coupling constants are diagnostic with the more shielded phosphinidene resonances and larger coupling constants being typical for the *E* isomers. Rh complexes Cp*(L)M=P have been synthesized with PPh_3 and PMe_3 ligands and Mes* (**57**) and Is as phosphinidene substituents [103]. One Co complex **58** with a PPh_3 ligand has been isolated, while a ^{31}P NMR resonance at δ 1047 ppm has been attributed to Cp(CO)Co=PMes* [103]. Theoretical calculations on geometries, ^{31}P NMR chemical shifts and J_{PP} coupling constants for the Co, Rh, and Ir phosphinidene complexes *E*/*Z*-Cp(L)M=PH (L=PH_3, CO) showed good agreement with those observed for the experimental structures [103]. Four osmium phosphinidene complexes Ar(L)Os=PMes* (Ar=benzene, *p*Cy; L=PPh_3, CO) have been synthesized with *p*Cy(CO)Os=PMes* as a mixture of *E*/*Z*-isomers [104]. Likewise four stable ruthenium complexes Ar(L)Ru=PMes* (Ar=benzene, *p*Cy; L=PPh_3, *t*-BuNC) have been isolated while (η^6-C_6H_6)(CO)Ru=PMes* was only identified by its low field ^{31}P NMR resonance at δ 896 ppm [104]. Crystal structures were reported for **59** and **60**.

In addition, W and Ta complexed phosphinidenes have been synthesized with ligands other than the Cp or Cp* group. For example, Cowley et al. [105] obtained stable W-complex **61** from reacting $WCl_2(PMePh_2)_4$ with Ar*P=C=O and a variety of Ta-complexes **62** (with various R groups) were reported by Schrock and coworkers [106] that contain the tetradentate triamidoamine ligand $(Me_3SiNCH_2CH_2)_3N^{3-}$. Related Ta-complex **63**, having instead three *tert*-Bu_3SiO groups, was synthesized by Wolczanski and coworkers [107]. These three complexes **61–63** are linear M=P-R phosphinidenes with short M=P bonds and contrast, e.g., the bent M=P-R forms of **52–60** (Table 5); **63** has also a distinctly shorter W=P bond length than **53**. The difference is also reflected in the much more shielded ^{31}P NMR chemical shift of the linear phosphinidene as compared to bent ones that can have very low field resonances. A representative example of this is found in the $\delta(^{31}P)$ of 970 ppm for Ni complex **64** that was reported recently by Hillhouse and coworkers [108]. Related cationic Ni complex $[(Ph_3P)_3Ni{=}P{=}CH_2]^+AlCl_4^-$, reported earlier by Niecke and coworkers, has a ^{31}P NMR resonance at 504 ppm (J_{PP}=35 Hz) and is assumed to posses P=C double bond character [109].

Extremely little chemistry has been explored with these nucleophilic phosphinidene complexes. Schrock and coworkers showed that the R substituent of **62** can be exchanged with Li/RX, presumably via the "$[N_3N]Ta{=}PLi$", and that these react with pivaldehyde to give *trans*-phosphaalkenes [106b]. Hey-Hawkins and coworkers reported that acetone inserts into the Ta=P bond of $Cp^*Cl_2Ta{=}PIs$ to give $Cp^*Cl_2Ta\{(OCMe_2)_2PIs\}$ [98]. Of all the nucleophilic phosphinidenes Zr complex **53** is the most extensively studied one [110]. A compilation of its reactions is given in the 'wagon wheel'. The most prominent ones are the halide displacement reactions of organic halides that enable access to a spectrum of phosphiranes, phosphirenes, and phospholenes [93c]. Four-membered ring structures $(PSn)_2$ and $(PGe)_2$ are obtained from dialkyl tin and germanium dihalides, respectively, while in the absence of trapping reagents or with less bulky substituents the $(ZrP)_2$ dimer is formed [93c, 111]. Reactions with aldehydes result in phosphaalkenes. Interestingly, **53** reacts with alkynes, but not with olefins, to give a four-membered metallocycle presumably formed via initial π-complexation of the alkyne to the transition metal center [112].

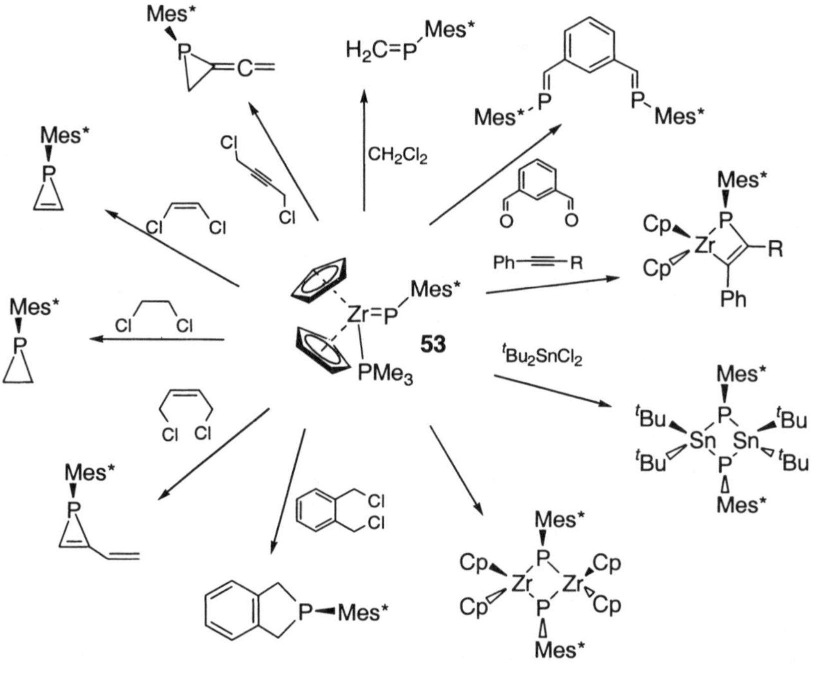

The group of Protasiewicz, who has reported Cp_2Zr complexed phosphinidenes with very bulky substituents and phosphine ligands, has explored the interchange of ligands and the formation of diphosphenes [113].

The reactivity of the $Cp^*(L)Ir{=}PMes^*$ ($L{=}PPh_3$, CO) phosphinidene complexes is much less diverse than those with Zr. Only the formation of phosphaalkenes has been observed in the reaction with CH_2I_2 and CHI_3 [102]. This reduced reactivity of the Ir complexes as compared to Zr complex **53** has been attributed to the much stronger binding of the 'additional' stabilizing phosphine ligand [103]. The convenient interchange of the phosphine ligand and the formation of the Zr_2P_2 dimer from **53** is illustrative.

Dehydrohalogenation of $(Ar)MCl_2(Mes^*PH_2)$ has also been explored in the absence of a stabilizing ligand and shown to result in the formation of **66** in the case of the Ru complex, presumable via the Ru_2P_2 dimer **65** [104]. The formation of **68** was shown to result when the dehydrohalogenation was executed in the presence of 2-butyne, which has metallocycle **67** as likely intermediate [104].

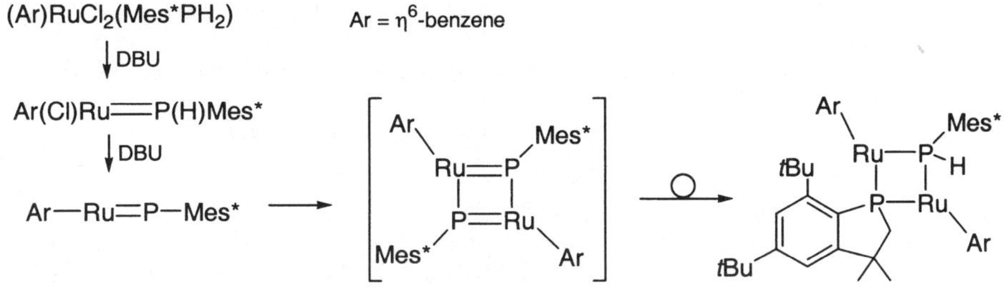

65 **66**

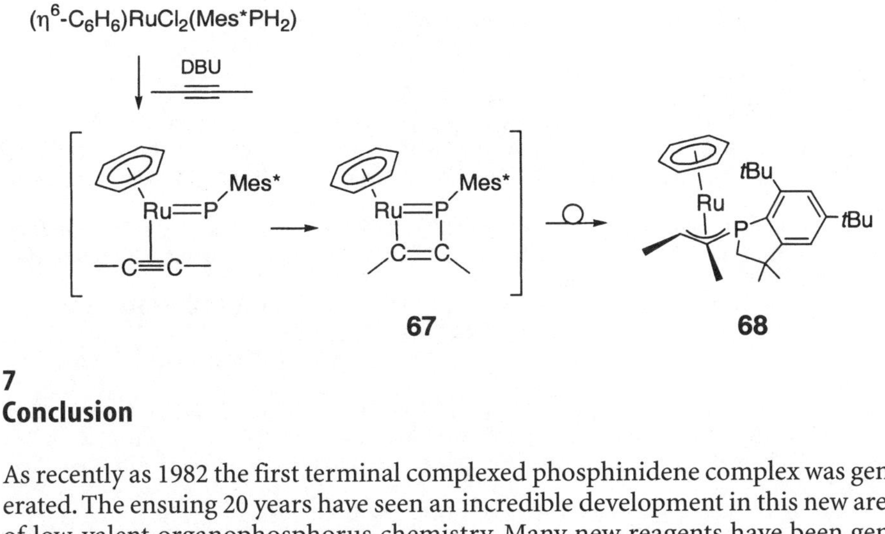

67 **68**

7 Conclusion

As recently as 1982 the first terminal complexed phosphinidene complex was generated. The ensuing 20 years have seen an incredible development in this new area of low-valent organophosphorus chemistry. Many new reagents have been generated. The electrophilic ones are the most reactive. A very diverse spectrum of applications has evolved resulting in many hitherto difficult to access organophosphorus compounds. The mechanistic pathways by which they are formed relate to those of the hydrocarbons, which is due to the diagonal relationship of carbon and phosphorus. More recently, much progress has been made in developing the chemistry of nucleophilic phosphinidenes. However, the exploration of their reactivity is still trailing. It should be evident though that many new discoveries are still to be made in this still rather young area of reactive intermediates.

8 References

1. The less used IUPAC recommended name is phosphanylidene
2. For an early review, see Mathey F (1987) Angew Chem 99:285; Mathey F (1987) Angew Chem Int Ed Engl 26:275
3. Mathey F, Tran Huy NH, Marinetti A (2001) Helv Chim Acta 84:2938
4. Lammertsma K, Vlaar MJM (2002) Eur J Org Chem 1127
5. Cowley AH (1997) Acc Chem Res 30:445; Cowley AH, Barron AR (1988) Acc Chem Res 21:88
6. Mathey F (1998) In: Dillon KB, Mathey F, Nixon JF (eds) Phosphorus: the carbon copy. Wiley, Chichester (UK)
7. Regitz M, Scherer OJ (eds) (1990) Multiple bonds and low coordination in phosphorus chemistry. Thieme, Stuttgart
8. Mathey F (ed) (2001) Phosphorus-carbon heterocyclic chemistry: the rise of a new domain. Elsevier
9. Mathey F (1990) Chem Rev 90:997
10. Arduengo AJ (1999) Acc Chem Res 32:913
11. Buron C, Guerret O, Gabbaï FP, Bertrand G (2000) Chem Rev 100:39; Solé S, Gornitzka H, Schoeller WW, Bourissou D, Bertrand G (2001) Science 292:1901; Kato T, Gornitzka H, Baceiredo A, Savin A, Bertrand G (2000) J Am Chem Soc 122:998; Buron C, Gornitzka H, Romanenko V, Bertrand G (2000) Science 288: 834
12. Haaf M, Schmedake TA, West R (2000) Acc Chem Res 33:704; Raabe G, Michl J (1985) Chem Rev 85:419
13. Frison G, Mathey F, Sevin A (1998) J Organometal Chem 570:225
14. Gaspar PP, Xiao M, Pae DH, Berger DJ, Haile TH, Chen T, Lei D, Winchester WR, Jiang P (2002) J Organomet Chem 646:68
15. Zittel PF, Lineberger WC (1976) J Phys Chem 65:1236
16. Nguyen MT, Van Keer A, Vanquickenborne LG (1996) J Org Chem 61:7077
17. Experimental: Berkowitz J, Greene JP, Cho H,Ruščić B (1987) J Chem Soc 86:1235. Theory: Yamaguchi Y, Van Huis TJ, Sherrill CD, Schaefer HF III (1997) Theor Chem Acc 97:341
18. Experimental: Leopold DG, Murray KK, Miller AES, Lineberger WC (1985) J Chem Phys 83:4849; Jensen P, Bunker PR (1988) J Chem Phys 89:1327. Theory: Yamaguchi Y, Schaefer HF III (1997) Chem Phys 225:23
19. Experimental: Engelking PC, Lineberger WC (1976) J Chem Phys 65:4323. Theory: Alexander MH, Werner HJ, Dagdigian PJ (1988) J Chem Phys 89:1388
20. Borden WT, Gritsan NP, Hadad CM, Karney WL, Kemnitz CR, Platz MS (2000) Acc Chem Res 33:765
21. Ehlers AW, Lammertsma K, Baerends EJ (1998) Organometallics 17:2738
22. Hamilton TP, Willis AG, Williams SD (1995) Chem Phys Lett 246:59
23. Li X, Weissman SI, Lin TS, Gaspar PP, Cowley AH, Smirnov AI (1994) J Am Chem Soc 116:7899
24. Li X, Lei D, Chiang MY, Gaspar PP (1992) J Am Chem Soc 114:8526
25. Cowley AH, Gabbaï F, Schluter R, Atwood D (1992) J Am Chem Soc 114:3142
26. Yoshifuji M, Sato T, Inamoto N (1988) Chem Lett 1735
27. Rahmoune M, Ko YYCYL, Carrié R, Tonnard F (1989) New J Chem 13:891
28. Creve S, Pierloot K, Nguyen MT (1998) Chem Phys Lett 285:429; Creve S, Pierloot K, Nguyen MT, Vanquickenborne LG (1999) Eur J Inorg Chem 107
29. Moss RA (1989) Acc Chem Res 22:15
30. Gonbeau D, Phister-Guillouzo G (1987) Inorg Chem 26:1799
31. Sudhakar PV, Lammertsma K (1991) J Org Chem 56:6067
32. Yates BF, Bouma WJ, Radom L (1987) J Am Chem Soc 109:2250
33. Ragavachari K, Chandrasekhar J, Gordon MS, Dykema KJ (1984) J Am Chem Soc 106:5853

34. Sudhakar PV, Lammertsma K (1991) J Am Chem Soc 113:5219
35. Sudhakar PV, Lammertsma K (1991) J Am Chem Soc 113:1899
36. Fritz G, Vaahs T, Fleischer H, Matern E (1989) Z Anorg Allg Chem 570:54; Fritz G, Vaahs T, Fleischer H, Matern E (1989) Angew Chem 101:324; Fritz G, Vaahs T, Fleischer H, Matern E (1989) Angew Chem Int Ed Engl 28:315
37. Shah S, Protasiewicz JD (1998) J Chem Soc Chem Commun 1585
38. Yoshifuji M, Shima I, Inamoto N, Hirotsu K, Higuchi T (1981) J Am Chem Soc 103:4587
39. Smith RC, Shah S, Protasiewicz JD (2002) J Organomet Chem 646:255; Smith RC, Ren T, Protasiewicz JD (2002) Eur J Inorg Chem 2779
40. Shah S, Protasiewicz JD (2000) Coord Chem Rev 210:181
41. Shah S, Yap GPA, Protasiewicz JD (2000) J Organomet Chem 608:12
42. Schmidpeter A (1990) In: Regitz M, Scherer OJ (eds) Multiple bonds and low coordination in phosphorus chemistry. Thieme, Stuttgart
43. Shah S, Simpson MC, Smith RC, Protasiewicz JD (2001) J Am Chem Soc 123:6925
44. Smith RC, Urnezius E, Lam KC, Rheingold AL, Protasiewicz JD (2002) Inorg Chem 41:5296
45. Ehlers AW, Baerends EJ, Lammertsma K (2002) J Am Chem Soc 124:2831
46. Marinetti A, Mathey F, Fischer J, Mitschler A (1982) J Am Chem Soc 104:4484; Marinetti A, Mathey F, Fischer J, Mitschler A (1982) J Chem Soc Chem Commun 667
47. Marinetti A, Charrier C, Mathey F, Fischer J (1985) Organometalics 4:2134
48. Marinetti A, Mathey F (1984) Organometallics 3:456
49. Lammertsma K, Chand P, Yang SW, Hung JT (1988) Organometallics 7:1875
50. Hung JT, Lammertsma K (1992) Organometallics 11:4365
51. Wang B, Lammertsma K (1994) J Am Chem Soc 116:10,486
52. Tran Huy NG, Inubushi Y, Ricard L, Mathey F (1997) Organometallics 16:2506
53. McKee ML, Ehlers AW, Lammertsma K(2002) WATOC 2002, Lugano, SW
54. Vlaar MJM, de Kanter FJJ, Schakel M, Lutz M, Spek AL, Lammertsma K (2001) J Organomet Chem 617/618:311
55. Mercier F, Deschamps B, Mathey F (1989) J Am Chem Soc 111:9098
56. (a) Streubel R, Ostrowski A, Wilkens H, Ruthe F, Jeske J, Jones PG (1997) Angew Chem Int Ed 36:378; Streubel R, Ostrowski A, Wilkens H, Ruthe F, Jeske J, Jones PG (1997) Angew Chem 109:409; (b) Streubel R, Kusenberg A, Jeske J, Jones PG (1994) Angew Chem Int Ed 33:2427; Streubel R, Kusenberg A, Jeske J, Jones PG (1994) Angew Chem 106:2564. (c) Streubel R (2002) Coord Chem Rev 227:175
57. Cowley AH, Geerts RL, Nunn CM (1987) J Am Chem Soc 109:6523
58. Sterenberg BT, Udachin KA, Carty AJ (2001) Organometallics 20:2657
59. Sterenberg BT, Udachin KA, Carty AJ (2001) Organometallics 20:4463
60. Sterenberg BT, Carty AJ (2001) J Organomet Chem 617/618:696
61. Wit JBM, van Eijkel GT, Schakel M, Lammertsma K (2000) Tetrahedron 56:137
62. Wit JBM, van Eijkel GT, de Kanter FJJ, Schakel M, Ehlers AW, Lutz M, Spek AL, Lammertsma K (1999) Angew Chem Int Ed 38:2596; Wit JBM, van Eijkel GT, de Kanter FJJ, Schakel M, Ehlers AW, Lutz M, Spek AL, Lammertsma K (1999) Angew Chem 111:2716
63. Nakazawa H, Buhro WE, Bertrand G, Gladysz JA (1984) Inorg Chem 23:3431
64. Vlaar MJM, Valkier P, de Kanter FJJ, Schakel M, Ehlers AW, Spek AL, Lutz M, Lammertsma K (2001) Chem Eur J 7:3551
65. Vlaar MJM, Ehlers AW, de Kanter FJJ, Schakel M, Spek AL, Lutz M, Sigal N, Apeloig Y, Lammertsma K (2000) Angew Chem Int 39:4127; Vlaar MJM, Ehlers AW, de Kanter FJJ, Schakel M, Spek AL, Lutz M, Sigal N, Apeloig Y, Lammertsma K (2000) Angew Chem 112:4296
66. Streubel R, Neumann C (1999) J Chem Soc Chem Commun 499
67. Streubel R, Tran Huy NH, Mathey F (1993) Synthesis 763; Mercier F, Ricard L, Mathey F, Regitz F (1991) J Chem Soc Chem Commun 1305; Streubel R, Ernst L, Jeske J, Jones PG (1995) J Chem Soc Chem Commun 2113
68. Hung JT, Yang SW, Chand P, Gray GM, Lammertsma K (1994) J Am Chem Soc 116:10,966
69. Hung JT, Wang SW, Gray GM, Lammertsma K (1993) J Org Chem 58:6786
70. Lammertsma K, Wang B, Hung JT, Ehlers AW, Gray GM (1999) J Am Chem Soc 121:11,650

71. Krill S, Wang B, Hung JT, Horan CJ, Gray GM, Lammertsma K (1997) J Am Chem Soc 119:8432
72. Tran Huy NH, Ricard L, Mathey F (2001) J Organomet Chem 617/618:748
73. Tran Huy NH, Marinetti A, Ricard L, Mathey F (2001) Organometallics 20:593
74. Komen CMD, Horan CJ, Krill S, Gray GM, Lutz M, Spek AL, Ehlers AW, Lammertsma K (2000) J Am Chem Soc 112:12,507
75. Alcaraz JM, Svara J, Mathey F (1986) Nouveau J Chim 10:321
76. Lammertsma K, Hung JT, Chand P, Gray GM (1992) J Org Chem 57:6557
77. Wang B, Lake CH, Lammertsma K (1996) J Am Chem Soc 118:1690
78. van Eis MJ, de Kanter FJJ, de Wolf WH, Lammertsma K, Bickelhaupt F (2000) Tetrahedron 56:129; van Eis MJ, Nijbacker T, de Kanter FJJ, de Wolf WH, Lammertsma K, Bickelhaupt F (2000) J Am Chem Soc 122:3033
79. Bulo RE, Ehlers AW, Grimme S, Lammertsma K (2002) J Am Chem Soc 124:13,903
80. Vlaar MJM, Ehlers AW, de Kanter FJJ, Schakel M, Spek AL, Lammertsma K (2000) Angew Chem Int Ed 39:2943; Vlaar MJM, Ehlers AW, de Kanter FJJ, Schakel M, Spek AL, Lammertsma K (2000) Angew Chem 112:3071
81. Tran Huy NH, Ricard L, Mathey F (1999) JCS Dalton Trans 2409; Tran Huy NH, Ricard L, Mathey F (1997) Organometallics 16:4501; Wang B, Ngyen KA, Srinivas GN, Watkins CL, Menzer S, Spek AL, Lammertsma K (1999) Organometallics 18:796
82. Tran Huy NH, Ricard L, Mathey F (2001) Angew Chem Int Ed 40:1253; Tran Huy NH, Ricard L, Mathey F (2001) Angew Chem 113:1293
83. van Eis MJ, Komen CMD, de Kanter FJJ, de Wolf WH, Lammertsma K, Bickelhaupt F, Lutz M, Spek AL (1998) Angew Chem Int Ed 37:1547; van Eis MJ, Komen CMD, de Kanter FJJ, de Wolf WH, Lammertsma K, Bickelhaupt F, Lutz M, Spek AL (1998) Angew Chem 110:1656
84. van Eis MJ, Zappey H, de Kanter FJJ, de Wolf H, Lammertsma K, Bickelhaupt F (2000) J Am Chem Soc 122:3386
85. Svara J, Mathey F (1986) Organometallics 5:1159
86. Tran Huy NH, Ricard L, Mathey F (1998) New J Chem 75
87. Inubushi Y, Tran Huy NH, Mathey F (1996) J Chem Soc Chem Commun 1903
88. Vlaar MJM, Valkier P, Schakel M, Ehlers AW, Lammertsma K (2002) Eur J Org Chem 1797
89. Tran Huy NH, Marinetti A, Ricard L, Mathey F (1992) Organometallics 11:1438
90. Vlaar MJM, Ehlers AW, Schakel M, Clendenning SB, Nixon JF, Lutz M, Spek AL (2001) Angew Chem Int 40:4412; Vlaar MJM, Ehlers AW, Schakel M, Clendenning SB, Nixon JF, Lutz M, Spek AL (2001) Angew Chem 113:4524
91. Hitchcock PB, Lappert MF, Leung WP (1987) J Chem Soc Chem Commun 1282; Bohra R, Hitchcock PB, Lappert MF, Leung WP (1989) Polyhedron 8:1884
92. Niecke E, Hein J, Nieger M (1989) Organometallics 8:2290
93. (a) Hou Z, Breen TL, Stephan DW (1993) Organometallics 12:3158; (b) Hou Z, Stephan DW (1992) J Am Chem Soc 114:10,088; (c) Breen TL, Stephan DW (1995) J Am Chem Soc 117:11,914
94. Arney DSJ, Schnabel RC, Scott BC, Burns CJ (1996) J Am Chem Soc 118:6780
95. Vaughan GA, Hillhouse GL, Rheingold AL (1989) Organometallics 8:1760
96. Arif AM, Cowley AH, Nunn CM, Pakulski M (1987) J Chem Soc Chem Commun 994
97. Mahieu A, Igau A, Majoral JP (1995) Phosphorus Sulfur Silicon 104:235
98. Segerer U, Felsberg R, Blaurock S, Hadi GA, Hey-Hawkins E (1999) Phosphorus Sulfur Silicon 144/146:477
99. Hadi GAA, Fromm K, Blaurock S, Jelonek S, Hey-Hawkins E (1997) Polyhedron 16:721
100. Blaurock S, Hey-Hawkins E (2002) Eur J Inorg Chem 2975
101. Malisch W, Hirth UA, Grün K, Schmeußer M, Fey O, Weis U (1995) Angew Chem Int Ed 34:2500; Malisch W, Hirth UA, Grün K, Schmeußer M, Fey O, Weis U (1995) Angew Chem 107:2717
102. Termaten AT, Nijbacker T, Schakel M, Lutz M, Spek AL, Lammertsma K (2002) Organometallics 21:3196
103. Termaten AT, Aktas H, Schakel M, Ehlers AW, Lutz M, Spek AL, Lammertsma K (2003) Organometallics 22:1827

104. Termaten AT, Nijbacker T, Schakel M, Lutz M, Spek AL, Lammertsma K (2003) Chem Eur J 9:2200
105. Cowley AH, Pellerin B, Atwood JL, Bott SG (1990) J Am Chem Soc 112:6734
106. (a) Cummins CC, Schrock RR, Davis WM (1993) Angew Chem 105:758; Cummins CC, Schrock RR, Davis WM (1993) Angew Chem Int Ed 32:756; (b) Freundlich JS, Schrock RR, Davies WM (1996) J Am Chem Soc 118:3643
107. Bonanno JB, Wolczanski PT, Lobkovsky EB (1994) J Am Chem Soc 116:11,159
108. Melenkivitz R, Mindiola DJ, Hillhouse GL (2002) J Am Chem Soc 124:3846
109. David G, Niecke E, Nieger M, Radseck J, Schoeller WW (1994) J Am Chem Soc 116:2191
110. Stephan DW (2000) Angew Chem Int Ed 39:315; Stephan DW (2000) Angew Chem 112:322
111. Breen TL, Stephan DW (1996) Organometallics 15:4509
112. Breen TL, Stephan DW (1996) J Am Chem Soc 118:4204; Breen TL, Stephan DW (1996) Organometallics 15:5729
113. Urnezius E, Lam KC, Rheingold AL, Protasiewicz JD (2001) J Organomet Chem 630:193; Urnezius E, Shah S, Protasiewicz JD (1999) Phosphorus Sulfur Silicon 144/146:137

Top Curr Chem (2003) 229: 121 – 141
DOI 10.1007/b11153

New Vistas in Chemistry and Applications of Primary Phosphines

Kattesh V. Katti · Nagavarakishore Pillarsetty · Kannan Raghuraman

Departments of Chemistry, Physics and Radiology, University of Missouri-Columbia, Room 106, Allton Buildings, 301, Business Loop 70 (W), Columbia, MO 65211, USA
E-mail: kattik@health.missouri.edu

Primary phosphines (R–PH_2) are an important class of compounds in organophosphorus chemistry. Although discovered over a century ago, their chemistry and applications have gained prominence in recent years. This review discusses recent developments on synthesis, molecular structure, properties, and applications of primary phosphines. In particular, discussions on synthesis and properties emphasize recent results from our laboratory on the chemical architecture of amide, thioether, and carboxylate functionalized primary bisphosphines. The utility of bromo- and aminopropyl phosphines ($X(CH_2)_3PH_2$; X=Br or NH_2) as building blocks to produce 'designer' primary phosphines that display exceptional oxidative stability is described. The review also discusses the utility of carboxylate functionalized primary phosphines for incorporation on to peptides and their potential applications in catalysis and biomedicine.

Keywords. Primary phosphines, Thioether phosphines, Amido phosphines, Carboxylate phosphines, Functionalized phosphines

List of Abbreviations

CVD	Chemical Vapor Deposition
D-Lys	D-Lysine
Gly	Glycine
HBTU	*O*-Benzotriazol-1-yl-N,N,N′,N′-tetramethyluronium hexafluorophosphate
HMPB	1,2-Bis[hydroxymethyl phosphinobenzene]
HMPE	1,2-Bis[hydroxymethyl phosphinoethane]
HPLC	High Performance Liquid Chromatography
LHRH	Lutenizing Hormone Releasing Hormone
PTC	Phase transfer catalyst
SPPS	Solid Phase Peptide Synthesis
TFA	Trifluoro acetic acid

1 Introduction

Design and development of new chemical vectors that provide fundamental insights in chemical science, and also whose unique properties impact vital applications in multiple areas of catalysis, new materials and biopharmaceuticals, will be of paramount importance for the scientific growth and futuristic prospects of modern chemistry. One such class of compounds is primary phosphines ($R{-}PH_2$). The high reactivity of P–H bonds in primary phosphines with chemical reagents such as acid halides, carbonyl compounds, Michael acceptors, alkyl halides, halogens, and Lewis acids (e.g., boranes) etc. has been extensively utilized to produce a myriad of functionalized organophosphorus compounds [1, 2]. Specifically, phosphorus-carbon functionalized compounds obtained via transformations across P–H bonds are versatile in various applications that include: (a) ligands for use in the development of catalytically useful tertiary phosphines [3–8], (b) chelating frameworks for use in the development of new chemotherapeutic and radioimaging agents in nuclear medicine [9–11], (c) precursors for applications in chemical vapor deposition (CVD) processes [12–14], and (d) chemical precursors or synthons of antimicrobial and bactericidal agents [15]. Primary phosphines (RPH_2), especially alkyl phosphines, are, in general, air-sensitive compounds. Therefore, their backbone modification to afford new classes of alkyl-substituted phosphines are often challenging. Representative reactions of primary phosphines that result in a plethora of novel organophosphorus compounds are outlined in Scheme 1 [16–24]. Despite their versatile chemistry and important applications, synthetic and structural investigations on new chemical architectures of functionalized primary phosphines are limited [25–30].

We have a long standing interest in the design and development of heteroatom and organic groups functionalized primary phosphines [31–33]. In particular, in recent years we have addressed promising and versatile synthetic strategies for producing air/thermally stable multifunctional and multinuclear primary phos-

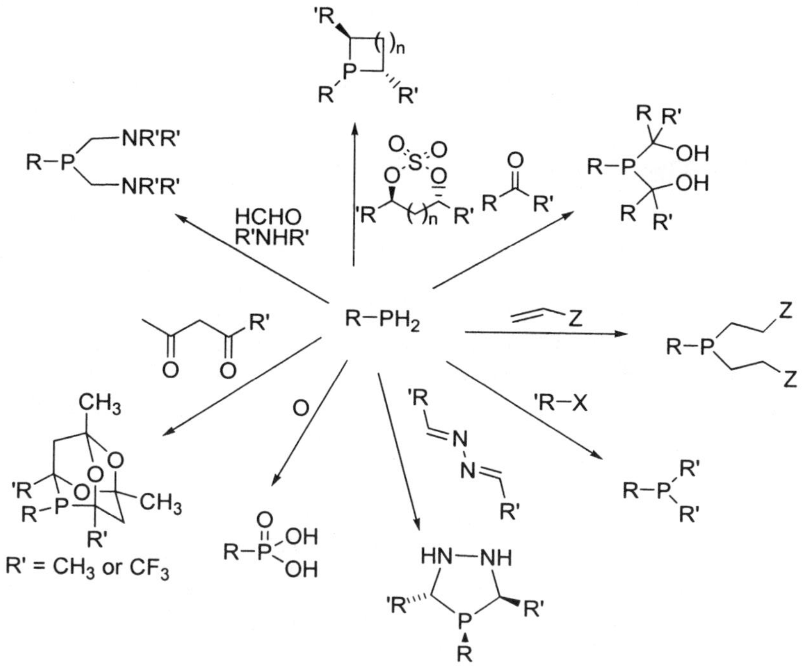

Scheme 1. Reaction versatility of P–H bonds in primary phosphines

phines and demonstrated their utility as effective synthons for the development of a new generation of water-soluble highly functionalized phosphines via formylation reactions of P–H bonds [31–33].

This review account will summarize latest research results on the design, development and properties of functionalized primary phosphines. In particular, the focus would be centered around recent results from our laboratory on the chemical architecture of heteroatom functionalized primary bisphosphines. We will also discuss synthetic protocols for the formylation reactions of functionalized primary phosphines to produce structurally diverse water-soluble hydroxymethyl phosphines. Finally, we will discuss the utility of carboxylate functionalized primary bisphosphines for incorporation on to peptides and their potential applications in catalysis and biomedicine.

2 Synthesis of Phosphines

2.1 Synthesis of Simple Alkyl/Aryl Primary Phosphines

Reduction of halophosphines [34–37] or alkyl phosphonates [38] using lithium aluminum hydride is commonly employed for the preparation of alkyl or aryl substituted primary phosphines (Eqs. 1–4):

$$2RPX_2 + LiAlH_4 \longrightarrow 2RPH_2 + LiAlX_4 \quad (1)$$

$$4RPOCl_2 + 3LiAlH_4 \longrightarrow 4RPH_2 + 2LiAlCl_4 + LiAl(OH)_4 \quad (2)$$

$$4RPO(OR')_2 + 3LiAlH_4 \longrightarrow 4RPH_2 + LiAl(OH)_4 + 2LiAl(OR')_4 \quad (3)$$

$$2RP(OR')_2 + LiAlH_4 \longrightarrow 2RPH_2 + LiAl(OR')_4 \quad (4)$$

Reactions of alkyl halides with alkali metal phosphides [39–41], addition reactions of olefins with PH_3 [42] or the pyrolytic cleavage of P–O bonds in RPO_2H_2 [43] are also reported for the preparation of alkyl/aryl functionalized primary phosphines (Eqs. 5–7):

$$MPH_2 + R\text{-}X \longrightarrow RPH_2 + MX \quad (5)$$

$$RCH{=}CH_2 + PH_3 \longrightarrow H_2PCHRCH_3 \text{ or } RCH_2CH_2PH_2 \quad (6)$$

$$3RPO_2H_2 \xrightarrow{\text{pyrolysis}} RPH_2 + 2RPO_3H_2 \quad (7)$$

Synthesis of primary phosphines via the direct alkali metallation of red phosphorus has also been reported (Eq. 8) [44]:

$$P_4 + 12Na + 8NH_4Br \xrightarrow{NH_3(liq)} 4NaPH_2 + 8NaBr + 8NH_3 \xrightarrow{RX} R\text{-}PH_2 \quad (8)$$

As shown in Eq. (9), sodium dithionite abstracts formaldehyde from hydroxymethyl phosphine ($R–P(CH_2OH)_2$) to produce a primary phosphine [26]:

$$R\text{-}P(CH_2OH)_2 \xrightarrow{NaS_2O_5\,(1\text{ eq.})} RPH_2 \quad (9)$$

Lithium aluminum hydride is a powerful reducing agent and often reduces carbonyl or other reduction susceptible functionalities present in the organophosphorus framework. Therefore, coexistence of $-PH_2$ and -COOH or $-CONH_2$ groups within the same molecule will require careful reaction strategies or the application of highly selective reducing agents as discussed in the following sections.

2.2
Amide Functionalized Primary Bisphosphines

Several different types of primary bisphosphines functionalized with mono- or diamides have been synthesized recently. The simplest among this series consists of a bisprimary phosphine appended to phenylamides as depicted in Scheme 2.

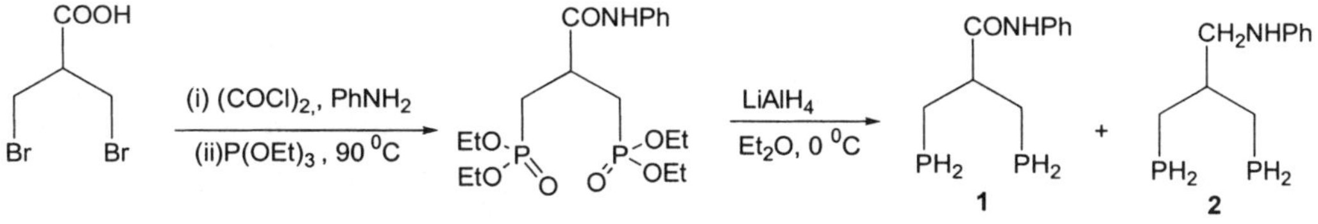

Scheme 2. Synthesis of bisprimary phosphines **1** and **2**

The bisphosphonate – upon reduction with lithiumaluminum hydride in ether at 0 °C – produced the amide functionalized primary bisphosphine (**1**) in good yields [45]. This reaction proceeded to reduce the amide group in **1** to produce the amine functionalized primary bisphosphine (**2**) in <5% yields. The amido bisprimary phosphine **1** is an air stable crystalline solid whereas the amine compound **2** is an oxidatively stable liquid. Separation of **1** and **2** in pure forms was achieved using column chromatography. The amidic bisprimary phosphine **1** was crystallized from chloroform and exhibits remarkable stability not only in the solid state but also in solution as well. The crystal structure of the air stable primary *bis*-phosphine **1** as shown in Fig. 1 is unprecedented to date.

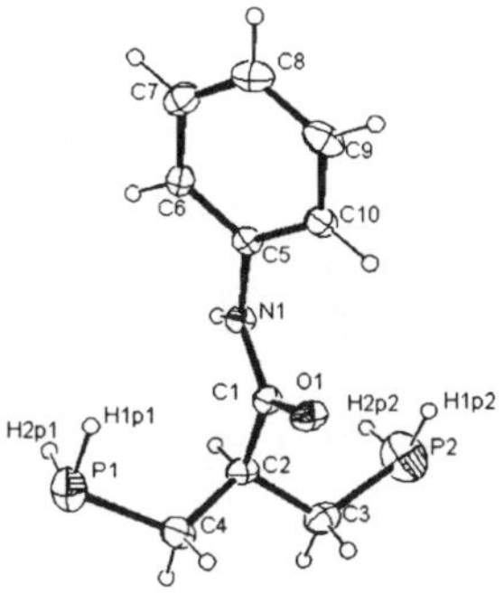

Fig. 1. Crystal structure of the amide functionalized bisprimary phosphine

Almost, all of the reported crystal structures of free primary phosphines have a bulky group attached to the phosphorus, thereby sterically crowding the phosphorus atom and rendering stability to P–H bonds [25–30]. In this context, the amide functionalized primary *bis*-phosphine **1** is the first example of its kind demonstrating excellent oxidative and thermal stability in the absence of any bulky substituent. The phosphine hydrogens in **1** did not show any hydrogen bonding with either the C=O or PH_2 groups. The P–C bond lengths are in the expected range of 1.80–1.89 Å, found in other primary phosphines [25–30]. Scheme 3.

Aminopropyl phosphine $H_2N(CH_2)_3PH_2$ **3**, synthesized via the $LiAlH_4$ reduction of the corresponding phosphonate, serves as an excellent building block for

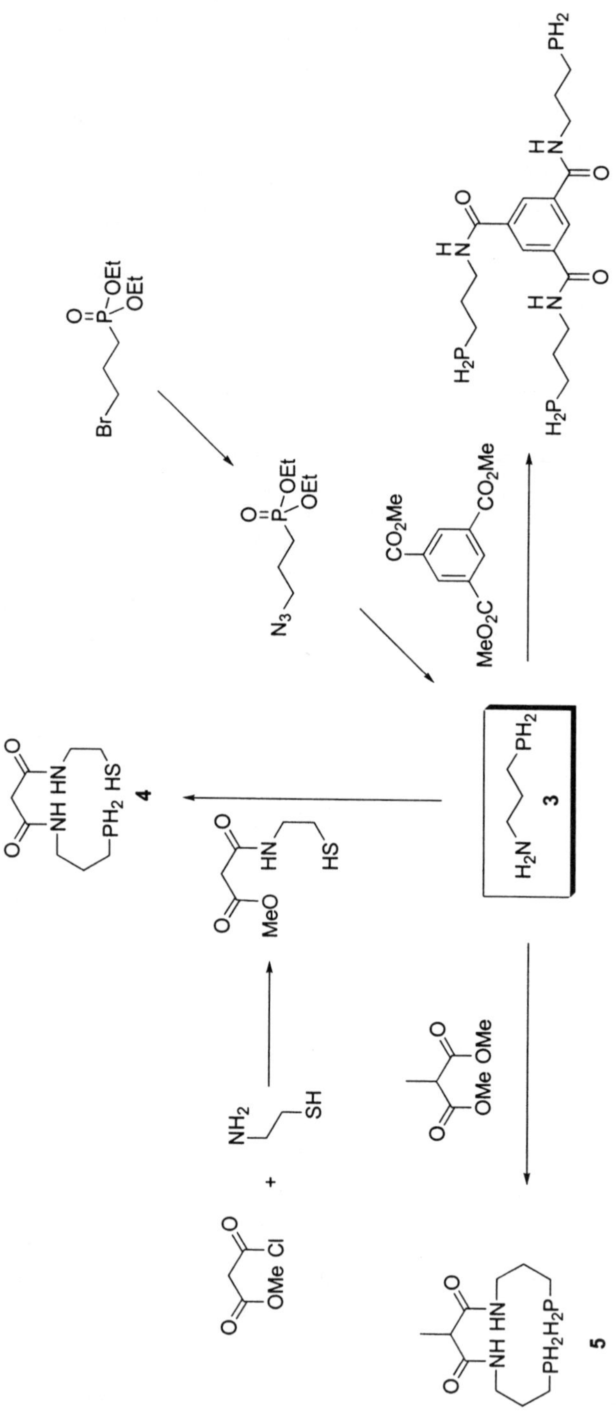

Scheme 3. Synthetic utility of aminopropyl phosphine **3**

the generation of novel amide functionalized bis/tris primary phosphines as outlined in Scheme 3 [46]. The reactions of aminopropyl phosphine **3** with functionalized organic esters produced a new generation of air stable thiol and amide functionalized (mono) **4**, (bis) **5**, and dendritic (tris) **6** primary phosphines in good yields (Scheme 3). The amidolysis of organic esters, as outlined in Scheme 3, exemplifies the rich chemistry of functionalized primary phosphines and also demonstrates the unique selectivity of the $-NH_2$ group in $H_2N(CH_2)_3PH_2$ (**3**) toward reactions with ester groups [46].

2.3 Thioether Functionalized Primary Bisphosphines

The thioether functionalized bisphosphonates, **7** and **8**, upon reduction with $LiAlH_4$ produced the corresponding S_2P_2 and S_3P_2 primary bisphosphine frameworks **9** and **10** respectively in good yields (Scheme 4) [47, 48].

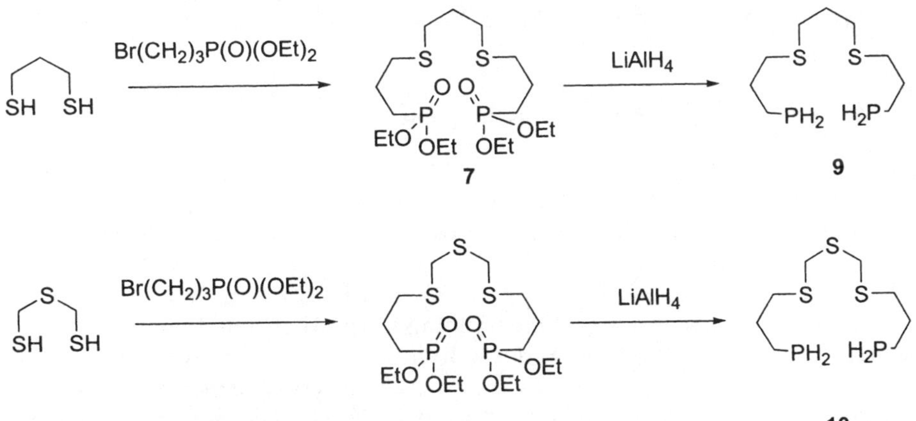

Scheme 4. Synthesis of thioether functionalized primary phosphines **9** and **10**

These thioether functionalized primary bisphosphines **9** and **10** showed modest oxidative stabilities and have found applications as novel precursors in the development of functionalized water-soluble phosphines via formylation reactions across P–H bonds (see below) [47].

Additional examples of thiol (**11**) [49] and thioether (**12**) functionalized primary phosphines are outlined in Scheme 5.

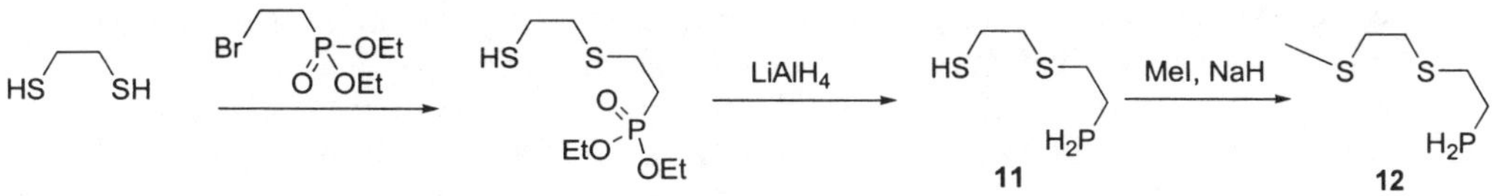

Scheme 5. Synthesis of thiol and thioether functionalized primary phosphines **11** and **12**

2.4
Mixed Phosphonate and Aromatic Amide Functionalized Primary Phosphines

The nucleophile assisted ring-opening reactions of phosphonate bearing phthalimide **13** has been utilized in the synthesis of mixed primary phosphine-phosphonate and aromatic amide functionalized primary bisphosphines as outlined in Scheme 6 [50].

For example, the interaction of aminopropyl phosphine (**3**) with the phthalimide **13** produced rare example of a phosphonate functionalized primary phosphine **14** in excellent yields (Scheme 6) [50]. The corresponding reaction of the phthalimide phosphonate **13** with 0.5 equivalent of hydrazine resulted in the bisphosphonate species **15** which upon further reduction with $LiAlH_4$ produced the primary bisphosphine **16** functionalized with aromatic amides. It is interesting to note that the reduction of the bisphosphine **15** with $LiAlH_4$ in dry THF at 0 °C was selective in terms of reducing only the phosphonate groups in **15**, although amide groups are susceptible for reduction under these conditions. The novel bisphosphine **16** was obtained in 80% yields, as an air stable, pale yellow solid [50].

2.5
Carboxylate Functionalized Primary Bisphosphines

Carboxylate, –COOH, and primary phosphine, $-PH_2$, groups represent the highly oxidized and reduced groups respectively within the chemical spectrum. Indeed, the coexistence of phosphanyl and carboxyl groups within the same molecule would be difficult because the reaction conditions that are used to reduce $-P(O)(OEt)_2$ group to a $-PH_2$ group can also reduce –COOH groups. In this context, preformed primary phosphine building blocks that contain reactive functionalities, such as halogens or amines, would be versatile for direct incorporation into carboxylated compounds. A schematic reaction sequence outlined in Scheme 7 provides a "nuts and bolts" strategy for the direct incorporation of carboxyl group within primary bisphosphine backbone.

This approach makes use of bromopropyl phosphine **17** as a key synthon obtained via the reduction of 3-bromopropyl phosphonate with dichloroaluminum hydride [10]. Reaction of bromopropyl phosphine **17** with the dianion of 6,8-dithiooctanoic acid produced the –COOH functionalized P_2S_2-primary bisphosphine framework **18** in >80% yields (Scheme 7) [10].

The application of 3-aminopropyl phosphine (**3**) [41, 46] as a building block for incorporation into –COOH functionalized frameworks provides an excellent example of the utility of preformed primary phosphine frameworks (Scheme 8) [46]. The reactions involved Michael addition of *tert*-butyl acrylate to malonic acid dimethyl ester to produce the intermediate adduct, 2-methoxycarbonyl-pentanedioc acid 5-*tert*-butyl ester 1-methyl ester, which upon treatment with trifluro-acetic acid (TFA) produced the corresponding diester acid, 2-methoxy-carbonyl-pentanedioic acid 1-methyl ester, in near quantitative yield. It is remarkable to note that the reaction of $NH_2(CH_2)_3PH_2$ (**3**) with the diester acid is highly selective as the –COOH group remained unattacked whereas the reaction occurred smoothly and selectively at the –COOMe groups to pro-

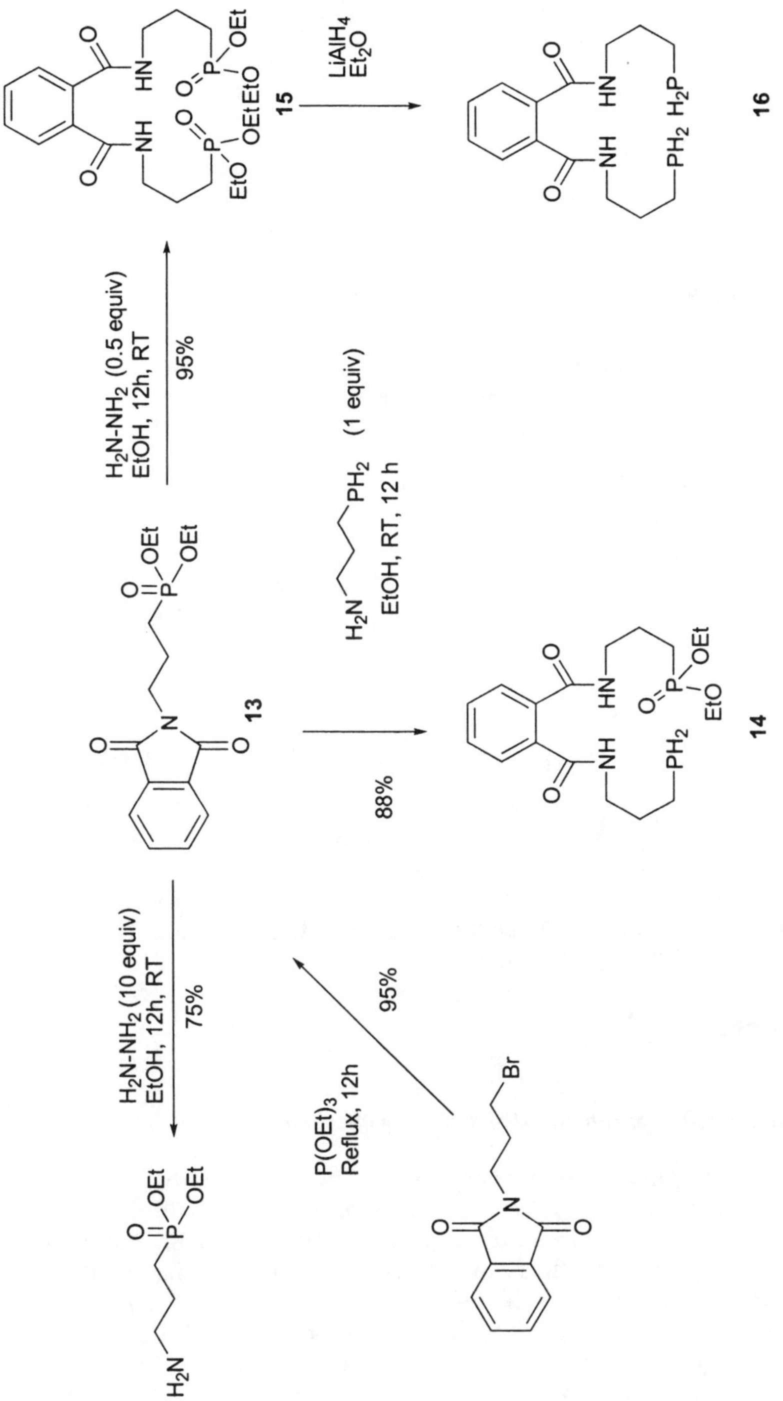

Scheme 6. Synthesis of amide functionalized primary phosphines

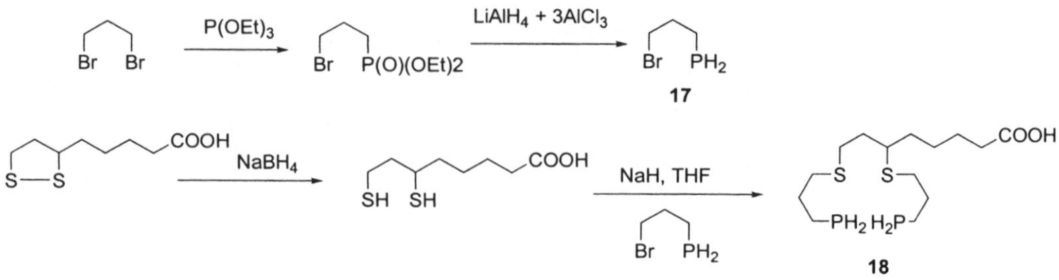

Scheme 7. Synthesis of carboxylate functionalized primary phosphine **18**

duce the novel carboxylate functionalized diamide bisprimary phosphine, 4,4-bis-(3-phosphanyl-propylcarbamoyl)butyric acid **19** in 51% yield (Scheme 8) [46].

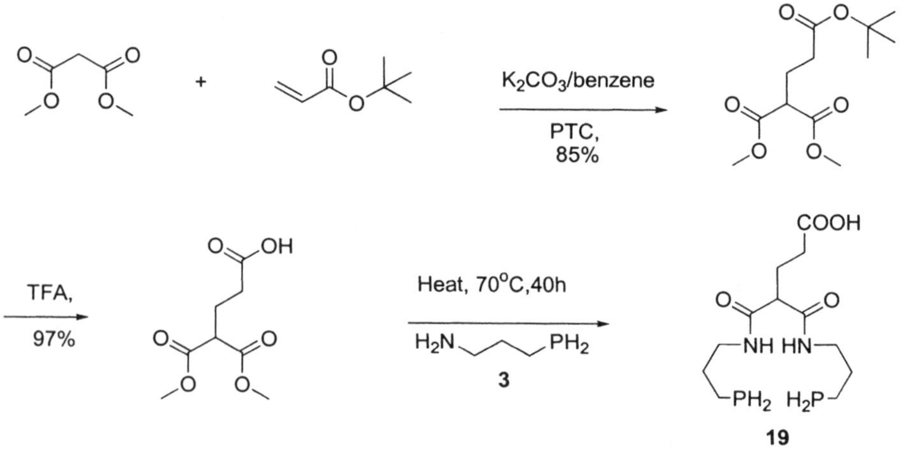

Scheme 8. Synthesis of carboxylate functionalized primary phosphine **19**

3 Properties

3.1 Oxidative Stability of Functionalized Primary Phosphines

Primary phosphines, in general, exhibit extreme hydrolytic thermal and oxidative instabilities. In fact, alkyl- mono (e.g., CH_2PH_2) or bis- (e.g., $H_2P(CH_2)_2PH_2$) phosphines are extremely dangerous during handling as they violently ignite in air. Efforts to improve oxidative/thermal stability have involved incorporating bulky alkyl/aryl substituents on the P^{III} center [25–27, 30]. For example, alkyl primary phosphines, as shown in Fig. 2, exhibit increased oxidative stability, presumably as a result of steric assistance of their bulky benzabarellene [25], iron dicyclopentadienyl [26], triptycyl [27], or a $^tBu_3C_6H_2$ [30] substituents (Fig. 2).

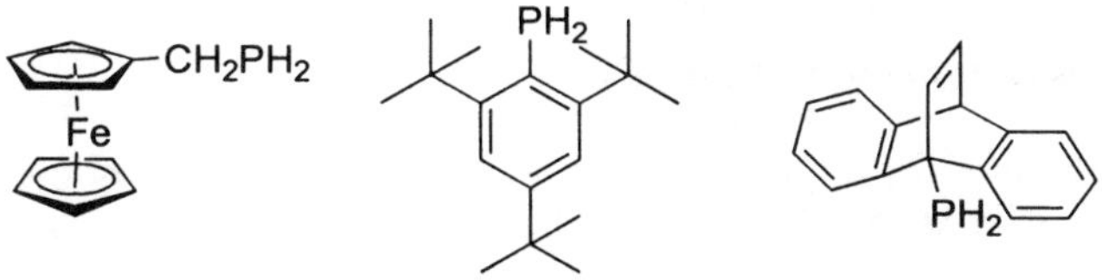

Fig. 2. Examples of air stable primary mono phosphines

It is striking to note that the new generation of amide, thioether and carboxylate functionalized primary bisphosphines, **1**, **10**, **16**, **18**, and **19** as shown in Fig. 3, provide examples of primary bisphosphines with unprecedented oxidative inertness.

In fact, the primary bisphosphines **1**, **10**, **16**, and **19** (Fig. 3) are air stable solids demonstrating exceptional oxidative stabilities. Recently, a primary bisphosphine **20** produced by dimerization reaction of anthracenyl primary phosphine has been shown to possess good oxidative stability [29].

Preliminary AM1 calculations carried out with the MOPAC program on **18** and related molecules suggest that there are atomic orbital contributions from the heteroatom (e.g., S in **18**) to the frontier molecular orbitals. It is conceivable, therefore, that there is negative hyperconjugation involving specific orbitals of S and the P^{III} centers in **18**. This electronic effect may explain the unusual stability towards oxidation of **18** and other heteroatom functionalized primary bisphosphines as described above [51].

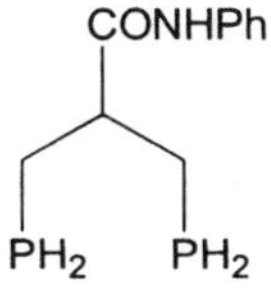

1

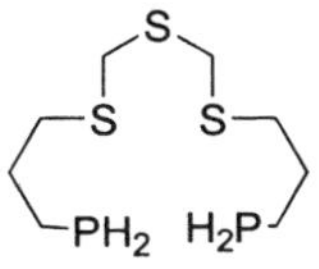

10

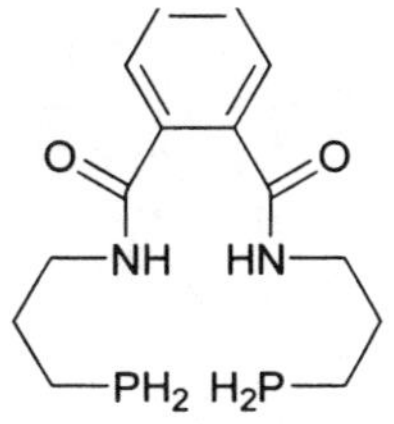

16

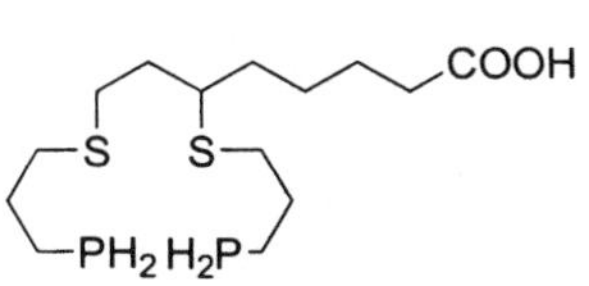

18

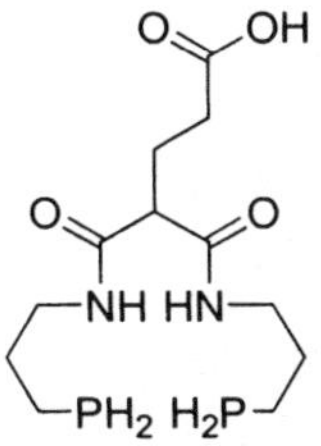

19

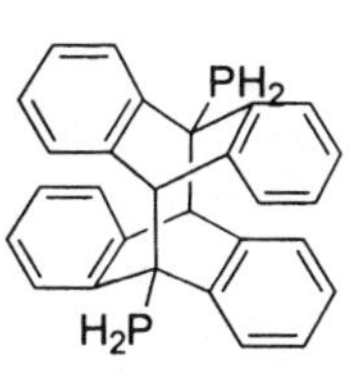

20

Fig. 3. Examples of air stable primary bis phosphines

3.2
Formylation Reactions of Primary Phosphines

The formylation of P–H bonds in mono and multiprimary phosphines, which result in the formation of hydroxymethyl phosphines, is among the facile useful reactions in organophosphorus chemistry. As shown in Scheme 9, formaldehyde in the presence of platinum catalysts transforms P–H bonds into hydroxymethyl (P–CH_2OH) functionalities (Scheme 9) [52].

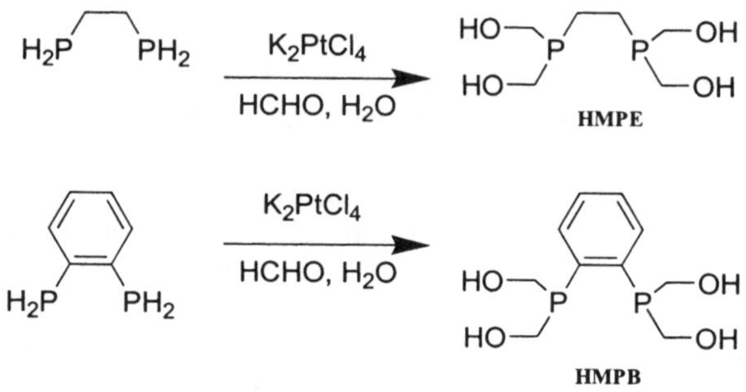

Scheme 9. Catalytic formylation of P–H bonds

Recent studies in our laboratory have demonstrated that formylation of P–H bonds can be achieved without the aid of transition metal catalysts under mild reaction conditions [47]. For example, amide and thioether functionalized primary phosphines, **5** and **9** respectively, upon treatment with 37% formaldehyde produced the corresponding amide/thioether functionalized water soluble phosphines **21** and **22** respectively in near quantitative yield (Scheme 10) [47].

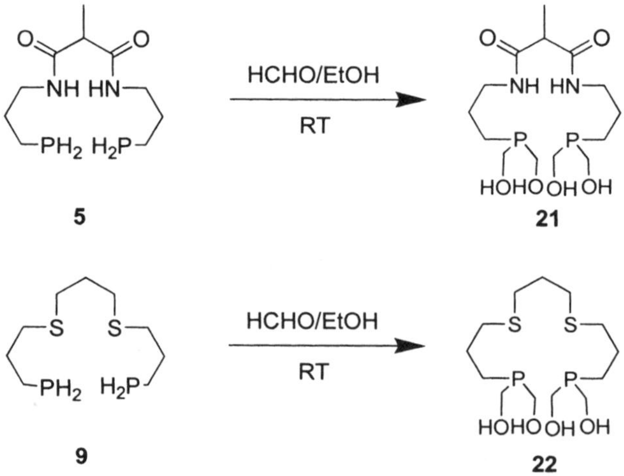

Scheme 10. Synthesis of hydroxymethyl phosphines from primary phosphines

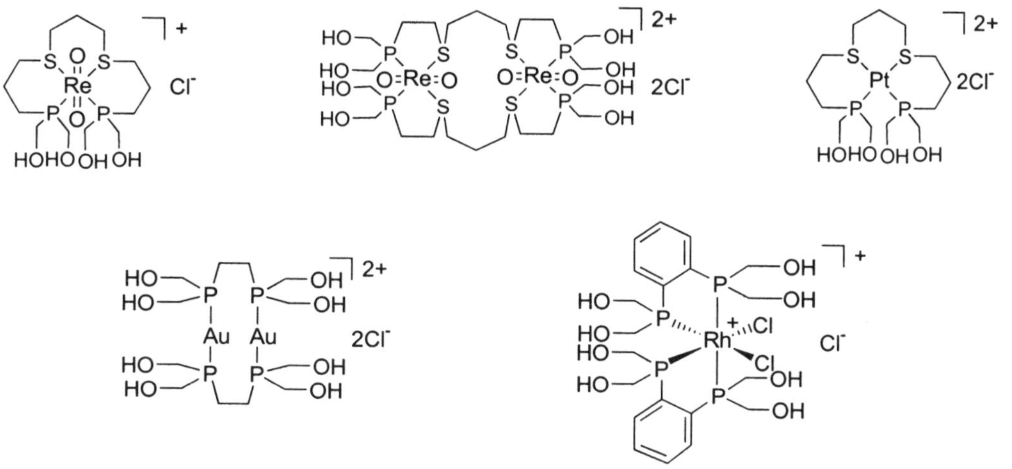

Fig. 4. Coordination modes of hydroxymethyl phosphines

Indeed, these reactions proceed at 25 °C in ethanol-aqueous media in the absence of transition metal catalysts. The ease with which P–H bonds in primary phosphines can be converted to P–C bonds, as shown in Schemes 9 and 10, demonstrates the importance of primary phosphines in the design and development of novel organophosphorus compounds. In particular, functionalized hydroxymethyl phosphines have become ubiquitous in the development of water-soluble transition metal/organometallic compounds for potential applications in biphasic aqueous-organic catalysis and also in transition metal based pharmaceutical development [53–62]. Extensive investigations on the coordination chemistry of hydroxymethyl phosphines have demonstrated unique stereospecific and kinetic propensity of this class of water-soluble phosphines [53–62]. Representative examples outlined in Fig. 4, depict bidentate and multidentate coordination modes and the unique kinetic propensity to stabilize various oxidation states of metal centers, such as Re(V), Rh(III), Pt(II) and Au(I), in aqueous media [53–62]. Therefore, the importance of functionalized primary phosphines in the development of multidentate water-soluble phosphines cannot be overemphasized.

4 Phosphorus NMR Spectroscopy

Phosphorus NMR spectroscopy serves as a valuable diagnostic tool in the characterization and structural elucidation of primary phosphine compounds. As shown in Table 1, RPH groups resonate in the –125 ppm to –150 ppm range with direct one bond ^{31}P–^{1}H coupling constants of 193–197 Hz.

It is interesting to note that, despite drastic changes in the chemical frameworks of primary bisphosphines, there are minimal/no differences in the chemical shifts and ^{31}P–^{1}H coupling constants (Table 1). The proton coupled ^{31}P NMR

Table 1. ^{31}P NMR spectroscopic data of primary phosphines

Compound number	Compound	$\delta\,^{31}$P	$^1J_{P-H}$ (Hz)
1	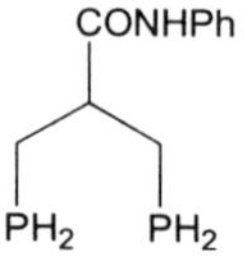	–143.0	197
2	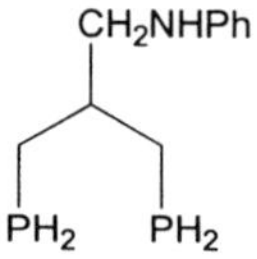	–149.6	197
3	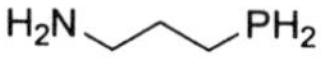	–135.3	194
4	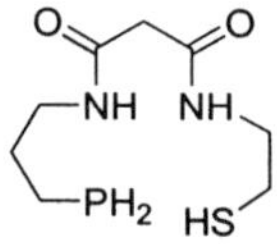	–134.6	195
5	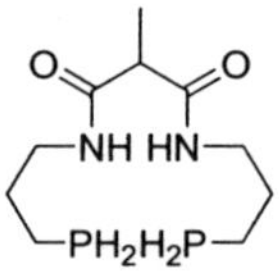	–135.6	194
6	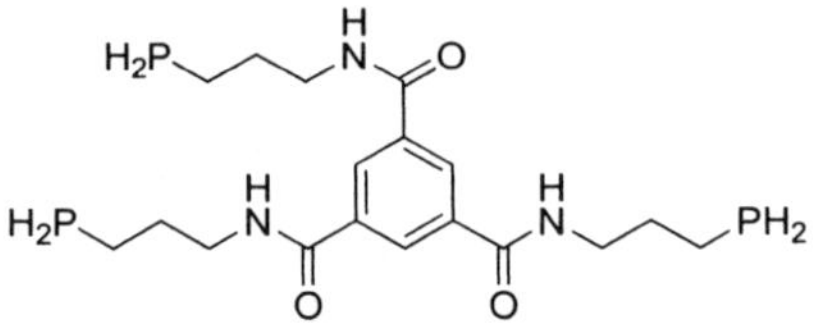	–135.6	195
9	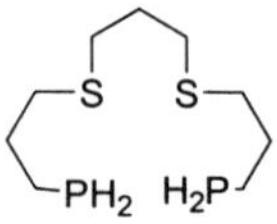	–137.5	192
10	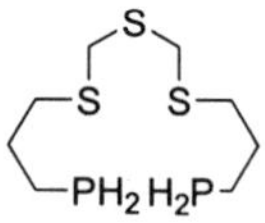	–136.3	194
11	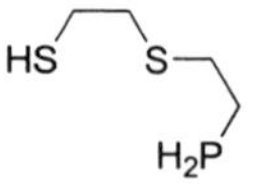	–140.6	195
12	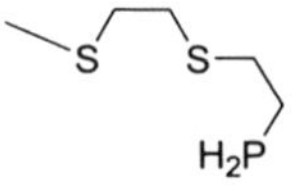	–141.1	192

Table 1 (continued)

Compound number	Compound	$\delta\,^{31}P$	$^{1}J_{P-H}$ (Hz)
14	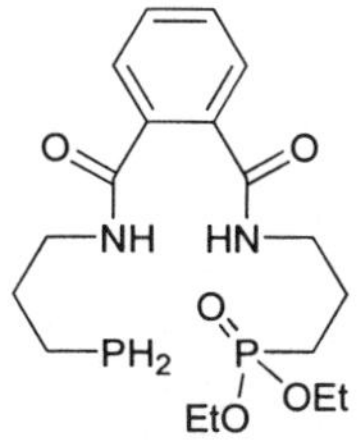	−135.1	194
16	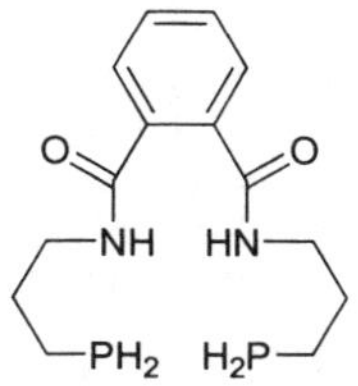	−135.6	193
17	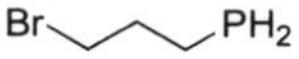	−138.4	–
18	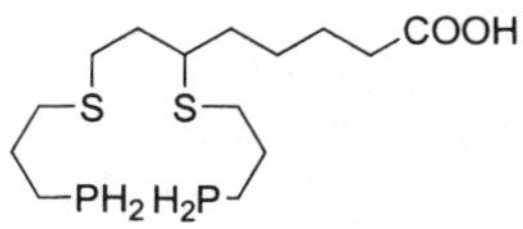	−136.3	193
19	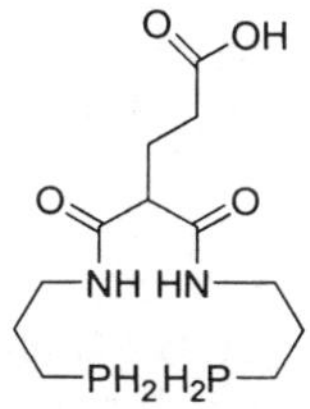	−135.6	194
23	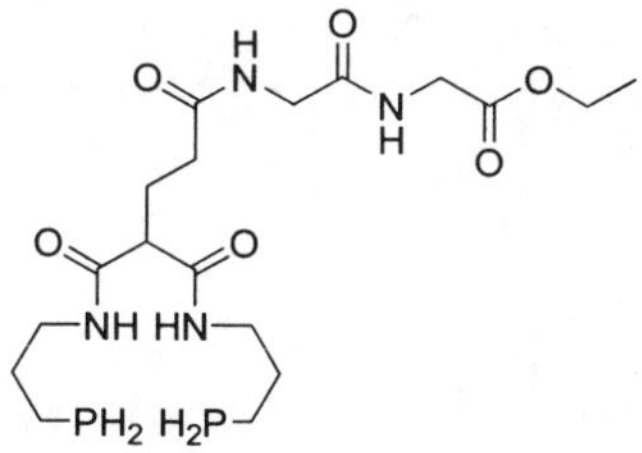	−135.9	194

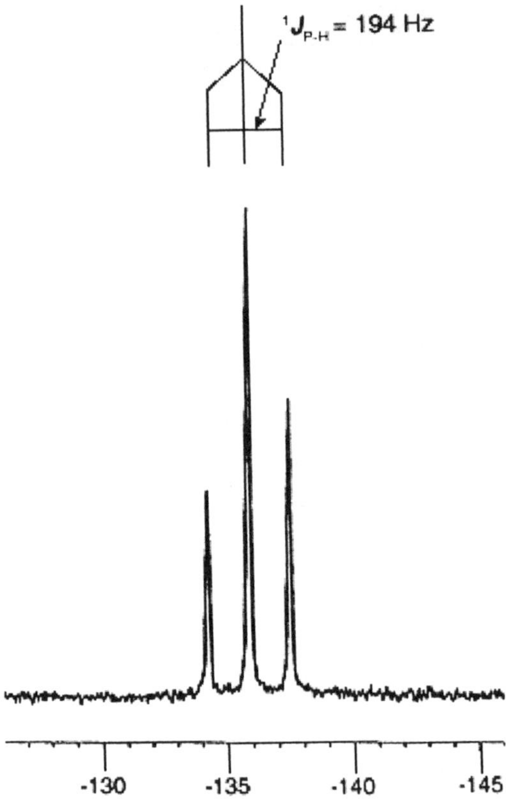

Fig. 5. Proton-coupled ^{31}P NMR spectrum of **23**

spectrum of a peptide-conjugated primary bisphosphine **23** as depicted in Fig. 5, shows the presence of unoxidized primary bisphosphine functionality. This figure exemplifies the diagnostic utility of the ^{31}P NMR spectroscopic technique to infer that the $-PH_2$ units are oxidatively inert even after subjecting the P_2N_2COOH primary phosphine precursor to harsh reaction conditions routinely used in the peptide synthesis [10, 46].

5
Biomedical and Catalytic Implications of Carboxylated Primary Bisphosphines

The carboxylate functionalized primary bisphosphines P_2S_2COOH **18** and P_2N_2COOH **19** (Schemes 7 and 8) provide new opportunities for use in catalytic and biomedical motifs. The carboxylate groups in **18** and **19** can be used to conjugate these phosphine ligating units on to peptides or proteins.

It is important to note that the secondary and tertiary structure of peptides to which they are attached may subsequently help in controlling the reactivity of phosphine-coordinated transition metals. Specifically, the chirality and related important stereospecific characteristics associated with biomolecules (e.g., peptides or proteins) may be transferred to the transition metals if peptides are immobilized with chelating units that are capable of coordinating with transition metals [63]. This approach of conjugating catalytically active transition metals to chiral biomolecules provides a straightforward synthetic route to harvest chiral compounds with potential applications in enantioselective catalysis. The incorporation of phosphines onto peptides will also help to engineer metal binding sites that may eventually provide conformational integrity, biospecificity, and enhanced enzymatic activities [64–68]. Additionally, bioconjugation of cytotoxic transition metals to receptor active peptides may eventually provide effective vehicles for delivering cytotoxic moieties to specific tumors through receptor-mediated agonist or antagonist interactions [69]. In this context, peptides (or receptor-binding biomolecules) containing phosphine substituents are important in the design and development of tumor-specific radiopharmaceuticals [70, 71]. Despite significant catalytic and biomedical applications offered by phosphines functionalized with peptides, synthetic strategies for producing such bioconjugates are still in their infancy. To demonstrate the feasibility of linking -COOH groups with peptides (and proteins), a synthetic protocol for linking compound **19** with a dipeptide, gly gly ethyl ester hydrochloride, has been developed (Scheme 11).

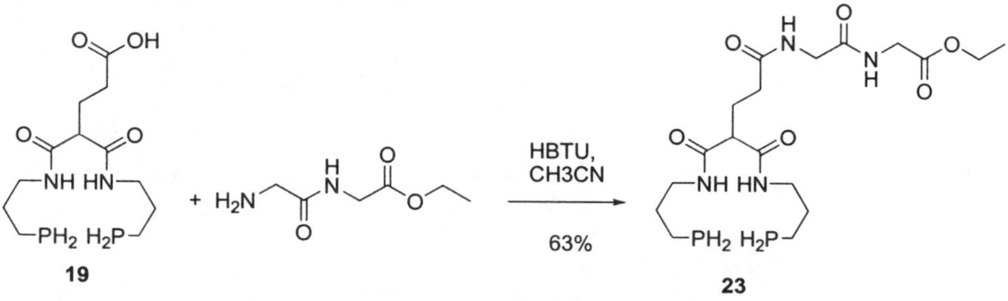

Scheme 11. Synthesis of peptide containing primary phosphine **23**

This method involved activation of the carboxylate of P_2N_2–COOH **19** using HBTU followed by interaction with the $-NH_2$ group of gly-gly peptide. The resulting P_2N_2 gly-gly peptide conjugate, {2-[4,4-bis-(3-phosphanyl-propylcarbamoyl)-butyrylamino]-acetylamino}-acetic acid ethyl ester **23**, was produced in 63% yields. The dipeptide conjugate **23** retained PH_2 units for direct complexation with catalytically or biologically useful metals.

To demonstrate the feasibility of linking P_2S_2–COOH compound **18** to biomolecules, a P_2S_2-D-Lys conjugate of a lutenizing hormone releasing hormone peptide, the D-Lys^6-LHRH conjugate **24**, was synthesized by automated solid-phase peptide synthesis (SPPS; Scheme 12). This method involved repeated use of variety of chemicals in high concentrations (including trifluroacetate (TFA)) for cleavage of peptide bound to the resin. Peptide conjugate **24** was purified by HPLC and analyzed by ^{31}P NMR spectroscopy and mass spectrometry. These data

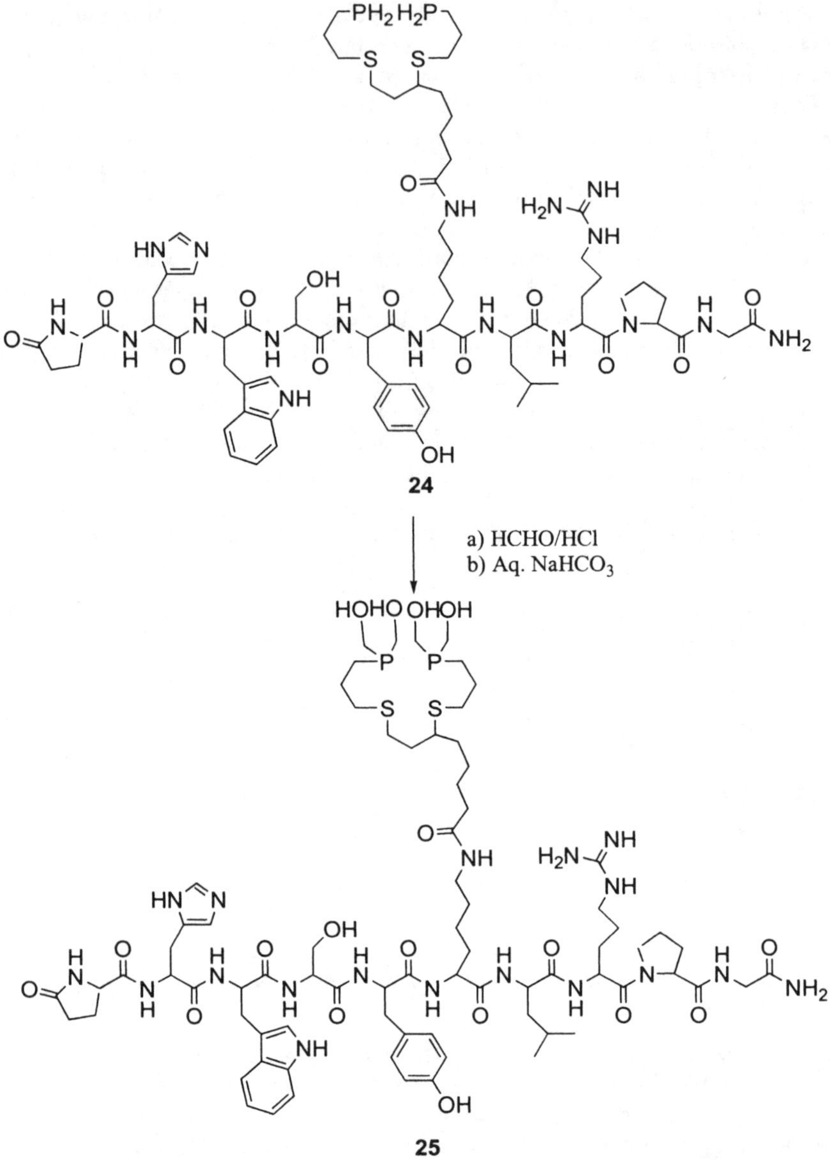

Scheme 12. Synthesis of peptide containing hydroxymethyl phosphane 25

demonstrate that the peptide conjugate **24** was formed in high yields with no modification of PH_2 groups. These results also confirm that the PH_2 groups of **18** and **19** are resistant to oxidation and are unreactive towards other functional groups in the peptides and the reagents used in SPPS [10]. The synthesis of such biomolecules which contain $-PH_2$ groups allows their conversion to hydrophilic alkyl phosphanes. Thus formaldehyde reacted rapidly with the PH_2 groups of **24** to produce the peptide-functionalized phosphane **25** (Scheme 12) via formylation of P–H bonds. Either PH_2 groups or their formylated analogues PR_2 may be used as a part of the chelator framework of biomolecules to form well-defined metalated conjugates by complexation with transition metals for potential catalytic or biomedical applications [10].

6 Concluding Remarks

The sequence of reactions summarized in Schemes 2–6 provides only a 'tip of iceberg' on the structural diversity and functional versatility of primary phosphines. Primary phosphines were once thought to be pyrophoric and highly unstable liquids. However, we have come a long way in producing 'designer' primary phosphines to suit the need of a specific chemical objective. The availability of new generations of primary phosphines will pave the way for future studies on comparisons in reactivity between 'P–H' and 'C–H' bonds by experimentalists and theoreticians. Unlike the carbon-hydrogen bonds, phosphorus hydrogen bonds display excellent reactivity toward both nucleophiles and electrophiles. It is this aspect of primary phosphines chemistry that makes them highly unique as chemical synthons for a variety of chemical transformations.

Functionally active preformed primary phosphines (e.g., $H_2N(CH_2)_3PH_2$ **3** or $Br(CH_2)_3PH_2$ **17**) will provide important building blocks to functionalize simple/complex molecules with primary phosphine functionalities. The 'user friendly' nature of the air stable primary bisphosphines (e.g., **1**, **10**, **16**, **18–20**) will open up new realms of exploratory research that utilize primary phosphines. It is also conceivable that the high oxidative stability and the ease with which primary phosphines can be incorporated on chiral backbones or peptides provide new opportunities for their applications in catalysis and biomedicine.

Acknowledgements. This work was supported by generous grants from the U.S. Department of Energy and DuPont-Merck Pharmaceuticals. Funding and facilities provided by the Departments of Radiology and Chemistry, and the Missouri University Research Reactor at the University of Missouri-Columbia are gratefully acknowledged. The work reported here would not have been possible without the painstaking efforts by various students, postdoctoral scientists, and technical staff whose names appear in the list of references.

7 References

1. Quin LD (2000) In: A guide to organophosphorus chemistry. Wiley-Interscience, New York
2. Corbridge DEC (1990) In: Phosphorus – an outline of its chemistry, biochemistry and technology. Elsevier Science, New York, chap 4
3. Brauer DJ, Hingst M, Kottsieper KW, Liek C, Nickel T, Tepper M, Stelzer O, Sheldrick WS (2002) J Organomet Chem 645:14
4. RajanBabu TV, Yan YY, Shin S (2001) J Am Chem Soc 123:10,207
5. Landis CR, Jin W, Owen JS, Clark TP (2001) Angew Chem Int Ed Engl 40:3432
6. Marinetti A, Jus S, Labrue F, Lemarchand A, Genet J-P, Ricard L (2001) Synthesis 14:2095
7. Burk MJ, Gross MF (1994) Tetrahedron Lett 35:9363

8. Burk MJ (1991) J Am Chem Soc 113:8518
9. Gali H, Hoffman TJ, Sieckman GL, Owen NK, Katti KV, Volkert WA (2001) Bioconjugate Chem 12:354
10. Gali H, Karra SR, Reddy VS, Katti KV (1999) Angew Chem Int Ed Engl 38:2020
11. Karra SR, Schibli R, Gali H, Katti KV, Hoffman TJ, Higginbotham C, Sieckman GL, Volkert WA (1999) Bioconjugate Chem 10:254
12. Scheper JT, Jayaratne KC, Liable-Sands LM, Yap GPA, Rheingold AL, Winter CH (1999) Inorg Chem 38:4354
13. Lewkebandara TS, Winter CH (1996) Adv Mater CVD 2:75
14. Lewkebandara TS, Proscia JW, Winter CH (1995) Chem Mater 7:1053
15. Randal F (1998) Todays Chem Work 7:34
16. Maier L (1972) In: Kosolapoff GM, Maier L (eds) Organic phosphorus compounds. Wiley, New York, p 1
17. Maier L (1963) Preparation and properties of primary, secondary and tertiary phosphines. In: Cotton FA (ed) Progress in inorganic chemistry. Interscience Publishers, New York, p 27
18. Berning DE, Katti KV, Barnes CL, Volkert WA (1999) J Am Chem Soc 121:1658
19. Gee V, Orpen AG, Phetmung H, Pringle PG, Pugh RI (1999) Chem Commun 901
20. Smith CJ, Reddy VS, Katti KV (1996) Chem Commun 2557
21. Ohff M, Holz J, Quirmbach M, Böerner A (1998) Synthesis 1391
22. Bourumeau K, Gaumont A-C, Denis J-M (1997) J Orgonomet Chem 529:205
23. Bourumeau K, Gaumont A-C, Denis J-M (1997) Tetrahedron Lett 38:1923
24. Petrov KA, Parshina VA (1968) Usp Khim 37:1218
25. Brynda M, Geoffroy M, Bernardinelli G (1999) Chem Commun 961
26. Goodwin NJ, Henderson W, Nicholson BK, Fawcett J, Russell DR (1999) J Chem Soc Dalton Trans 1785
27. Ramakrishnan G, Jouaiti A, Geoffroy M, Bernardinelli G (1996) J Phys Chem 100:10,861
28. Brauer DJ, Fischer J, Kucken S, Langhans KP, Stelzer O (1994) Z Naturforsch Teil B 49:1511
29. Wesemann J, Jones PG, Schomburg D, Heuer L, Schmutzler R (1992) Chem Ber 125:2187
30. Bartlett RA, Olmstead MM, Power PP, Sigel GA (1987) Inorg Chem 26:1941
31. Katti KV, Gali H, Smith CJ, Berning DE (1999) Acc Chem Res 32:9
32. Prabhu KR, Pillarsetty N, Gali H, Katti KV (2000) Current Sci 78:431
33. Katti KV (1996) Current Sci 70:215
34. Benneett FW, Emeléus HJ, Haszeldine RN (1954) J Chem Soc 3896
35. Evleth EM, Freeman LVD, Wagner RI (1962) J Org Chem 27:2192
36. Horvart RJ, Furst A (1952) J Am Chem Soc 74:562
37. Horner L, Beck P, Hoffmann H (1959) Chem Ber 92:2088
38. Maier L (1966) Helv Chim Acta 49:842
39. Wagner RI, Burg AB (1953) J Am Chem Soc 75:3879
40. Evers EC, Street EH Jr, Jung SL (1951) J Am Chem Soc 73:2038
41. Issleib K, Kuemmel R, Oehme H, Meissner I (1968) Chem Ber 101:3612
42. Stiles AR, Rust FF, Vaughan WE (1952) J Am Chem Soc 74:3282
43. Berrigan RA, Russell DK, Henderson W, Leach MT, Nicholson BK, Woodward G, Harris C (2001) New J Chem 25:322
44. Gusarova NK, Malysheva SF, Arbuzova SN, Trofimov BA (1998) Russ Chem Bull 47:1645
45. Pillarsetty N, Kandikere P, Gali H, Raghuraman K, Barnes C, Katti KV (manuscript in preparation)
46. Prabhu KR, Pillarsetty N, Gali H, Katti KV (2000) J Am Chem Soc 122:1554
47. Smith CJ, Reddy VS, Karra SR, Katti KV, Barbour LJ (1997) Inorg Chem 36:1786
48. Gali H (1999) Ph D thesis, University of Missouri-Columbia
49. Raghuraman K, Pillarsetty N, Katti KV, Barnes CL (manuscript in preparation)
50. Gali H, Prabhu KR, Karra SR, Katti KV (2000) J Org Chem 65:676
51. Gali H, Katti KV (unpublished results)
52. Reuter M, Orthner L (1958) German Patent 1,035,135; Chem Abstr (1960) 54:14,125a

53. Raghuraman K, Pillarsetty NK, Volkert WA, Barnes CL, Jurisson S, Katti KV (2002) J Am Chem Soc 124:7276
54. Reddy VS, Katti KV, Barnes CL (1995) Inorg Chim Acta 240:347
55. Reddy VS, Berning DE, Katti KV, Barnes CL, Volkert WA, Ketring AR (1996) Inorg Chem 35:1753
56. Reddy VS, Katti KV, Barnes CL (1996) J Chem Soc Dalton Trans 1301
57. Berning DE, Katti KV, Barnes CL, Volkert WA, Ketring AR (1997) Inorg Chem 36:2765
58. Berning DE, Katti KV, Barnes CL, Volkert WA (1997) Chem Ber 130:907
59. Smith CJ, Reddy SV, Katti KV (1998) J Chem Soc Dalton Trans 1365
60. Berning DE, Katti KV, Singh PR, Higginbotham C, Reddy VS, Volkert WA (1996) J Nucl Med Biol 23:617
61. Schibli R, Karra SR, Gali H, Katti KV, Higginbotham C, Volker WA (1998) Radiochim Acta 83:211
62. Smith CJ, Li N, Katti KV, Higginbotham C, Volkert WA (1997) Nucl Med Biol 24:685
63. Severin K, Bergs R, Beck W (1998) Angew Chem Int Ed 37:1634
64. Degrado SJ, Mizutani H, Hoveyda AH (2001) J Am Chem Soc 123:755
65. Gilbertson SR, Collibee SE, Agarkov A (2000) J Am Chem Soc 122:6522
66. Gilbertson SR, Wang X (1996) J Org Chem 61:434
67. Gilbertson SR, Wang X, Hoge GS, Klug CA, Schaefer J (1996) Organometallics 15:4678
68. Gilbertson SR, Chen G, McLoughlin M (1994) J Am Chem Soc 116:4481
69. Pinski J, Yano T, Miller G, Schally AV (1992) Prostate (NY) 20:213
70. Heeg MJ, Jurisson SS (1999) Acc Chem Res 32:1053
71. Volkert WA, Hoffman TJ (1999) Chem Rev 99:2269

Top Curr Chem (2003) 229: 143–159
DOI 10.1007/b11154

Imido Analogues of Phosphorus Oxo and Chalcogenido Anions

Tristram Chivers

Department of Chemistry, University of Calgary, Calgary, Alberta, T2N 1N4, Canada
E-mail: chivers@ucalgary.ca

This review deals with the chemistry and coordination complexes of isoelectronic analogues of common oxo-anions of phosphorus such as PO_3^-, PO_4^{3-}, RPO_3^{2-} and $R_2PO_2^-$. The article begins with a discussion of homoleptic systems in which all of the oxo ligands are replaced by imido (NR) groups. This is followed by an account of heteroleptic phosphorus-centered anions, including $[RN(E)P(\mu\text{-}NR')_2P(E)NR]^{2-}$, $[EP(NR)_3]^{3-}$, $[RP(E)(NR)_2]^{2-}$ and $[R_2P(E)(NR')]^-$ (E=O, S, Se, Te). The emphasis is on the wide variety of coordination modes exhibited by these polydentate ligands, which have both hard (NR) and soft (S, Se or Te) centers. Possible applications of their metal complexes include new catalytic systems, coordination polymers with unique properties, and novel porous materials.

Keywords. Anionic ligands, Phosphorus, Imido, Chalcogenido, Metal complexes

List of Abbreviations

p-cymene	1-Methyl, 4-ethylbenzene
naph	Naphthyl
THF	Tetrahydrofuran
TMEDA	Tetramethylethylenediamine

1 Introduction

Phosphorus forms a wide variety of oxoanions, which outnumber those formed by any other element except silicon [1]. Many of these anionic ligands are of industrial importance and their derivatives are essential components of numerous biological processes. Some of the more common examples include orthophosphate PO_4^{3-} (**1**), phosphite HPO_3^{2-} (**2**) and metaphosphate PO_3^- (**3**).

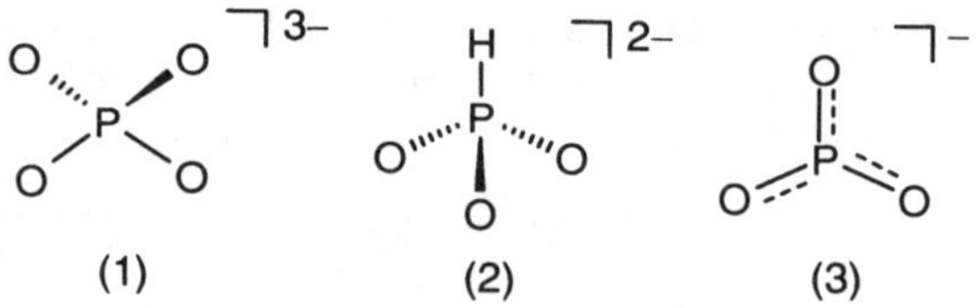

The imido group NR (R=H, alkyl, aryl) is isoelectronic with an oxo (O) ligand. This simple relationship raises the possibility of a related series of polyimido anions of phosphorus in which the oxo ligands are replaced by imido substituents. Partial replacement gives rise to heteroleptic imidooxo anions, for which sulfur or selenium analogues can be envisaged through the replacement of oxygen by a heavier chalcogen. Complete replacement generates homoleptic polyimido anions of phosphorus such as $[P(NR)_4]^{3-}$, $[HP(NR)_3]^{2-}$ and $[P(NR)_3]^-$.[a]

Since the mid-1990s investigations of polyimido anions of p-block elements have become an active area of main group chemistry [2, 3], and increasing attention is being accorded to phosphorus-centered anions[4]. The purpose of this overview is to survey these recent advances for both homoleptic and heteroleptic systems with emphasis on the conceptual thinking that has resulted in significant progress.

[a] Tris(imido)phosphonates $[R'P(NR)_3]^{2-}$ and bis(imido)phosphinates $[R'_2P(NR)_2]^-$ are not, strictly speaking, homoleptic anions. In this article, however, homoleptic refers to anions that contain only imido (NR) ligands, in addition to the H or R′ substituent directly attached to phosphorus (if any), whereas the term heteroleptic is used for anions that involve both imido and oxo (or chalcogenido) ligands.

2
Homoleptic Polyimido Anions of Phosphorus

2.1
Bis(imido)phosphorus(III) Anions $[P(NR)_2]^-$

From a historical perspective the first example of a polyimido anion of phosphorus, the 1,3-diaza-2-phosphaallyl anion $[P(NMes^*)_2]^-$ ($Mes^*=2,4,6\text{-}{}^tBu_3C_6H_2$), isoelectronic with the hypothetical PO_2^- anion was reported by Lappert et al. [5]. The lithium derivative $Li[P(NMes)_2^*]$ is prepared by lithiation of Mes*PN(H)PNMes* with Li^{n}Bu (Eq. 1), a procedure that has also been used for the synthesis of unsymmetrical derivatives $\{Li[P(NR)(NMes^*)]\}_2$ ($R=CPh_3$, 1-Ad, ^{t}Bu) [6]. In a similar manner the lithiation of *cis*-^{t}BuN(H)P(μ-N^{t}Bu)$_2$PN(H)^{t}Bu in toluene yields the tetramer $\{Li[P(N^tBu)_2]\}_4$ which contains the $[P(N^tBu)_2]^-$ anion [7]:

$$^nBuLi + Mes^*N{=}P(NHMes^*) \longrightarrow LiP(NMes^*)_2 + {}^nBuH \qquad (1)$$

The unsolvated complexes $\{Li[P(NR)(NR')]\}_n$ may adopt monomeric, dimeric or tetrameric structures as indicated by the examples (**4**) [7], (**5**) [6] and (**6**) [7]. Variable temperature NMR studies reveal that the tetramer (**6**) dissociates into the corresponding dimer in solution [7]. The dimeric cubic structure is stabilized in the THF-solvated complex $\{THF \cdot Li[P(N^tBu)_2]\}_2$ (**7**) [8].

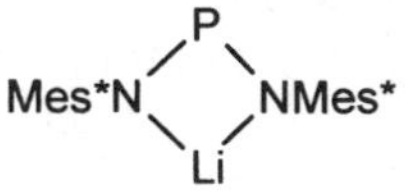

(4)

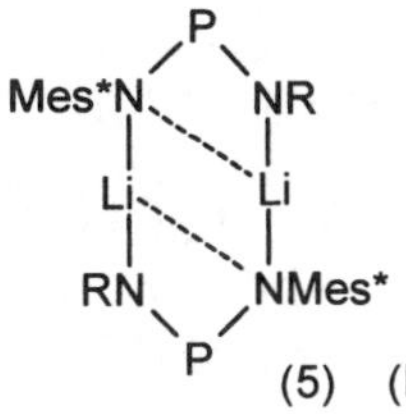

(5) (R = 1-Ad, ^{t}Bu)

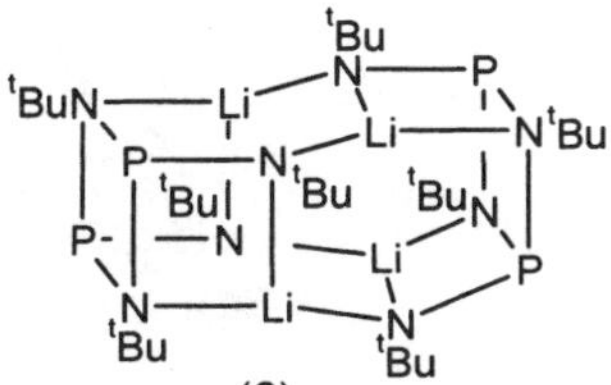

(6)

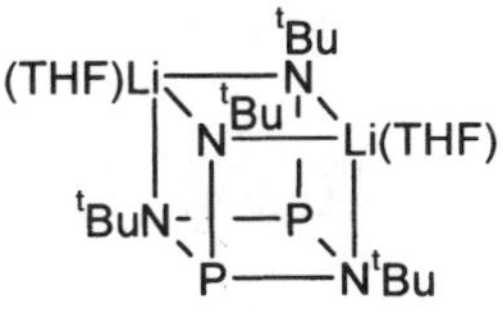

(7)

With the exception of a brief report of a dimethylaluminum complex [5], the coordination chemistry of the monomeric anion in (**4**) has not been investigated. By contrast, Stahl and co-workers have carried out extensive studies of both main group element and transition-metal complexes of the chelating dianion in the cube (**7**), which have been summarized in a recent review [9]. A noteworthy feature of the ligand behaviour of this *N,N'* chelating dianion is the additional involvement of a third (endocyclic) N^{t}Bu group in coordination (**8**) [10–13]. Another interesting observation is the involvement of the monomeric anion $[P(N^tBu)_2]^-$, generated by the cycloreversion of (**7**), in the formation of the spirocyclic gallium complex (**9**) [13].

Group 4 metal complexes of the dianion $[^tBuNP(\mu\text{-}N^tBu)_2PN^tBu]^{2-}$ polymerize ethylene in the presence of a co-catalyst, but they are readily deactivated [10, 14]. This behaviour is attributed to coordination of the lone-pair electrons on the phosphorus(III) centers to Lewis acid sites, which initiates ring opening of the ligand [15].

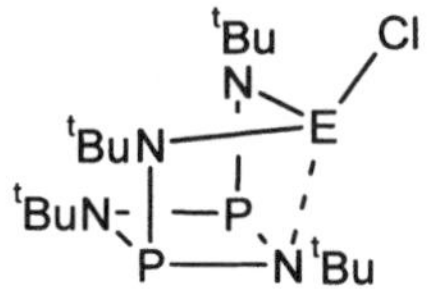

(8) E = Sb, Bi, Ga

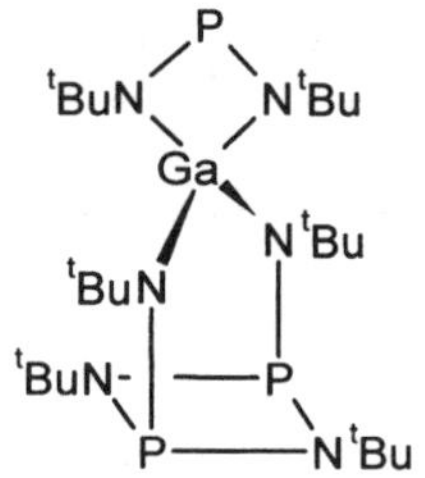

(9)

2.2 Tris(imido)metaphosphate Anions $[P(NR)_3]^-$

Niecke et al. have prepare polyimido analogues of the metaphosphate ion, PO_3^-, by lithiation of the corresponding amido compounds [16]. Thus the monomeric solvent-separated ion pair $[(THF)_4Li][P(NMes^*)_3]$ (**10**) is obtained by treatment of $(Mes^*N)_2P(NHMes^*)$ with ^nBuLi [16]. A monomeric contact ion pair (**11**) containing the unsymmetrical anion $[P(N^tBu)_2(NMes^*)]^-$ has also been reported [16]. By contrast the dilithium derivative of the trisimidometaphosphate $[P(N^tBu)_3]^-$ forms a dimer (**12**) [17], with a cubic structure reminiscent of that of (**7**).

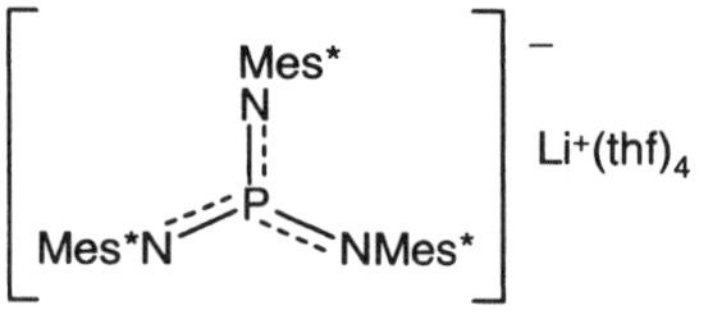

(10)

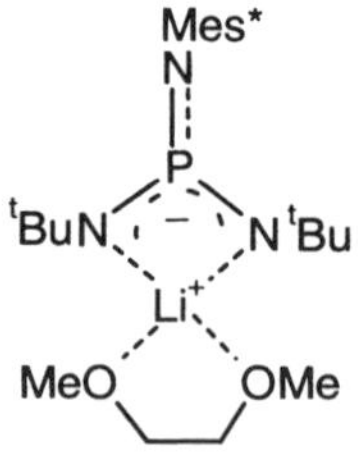

(11)

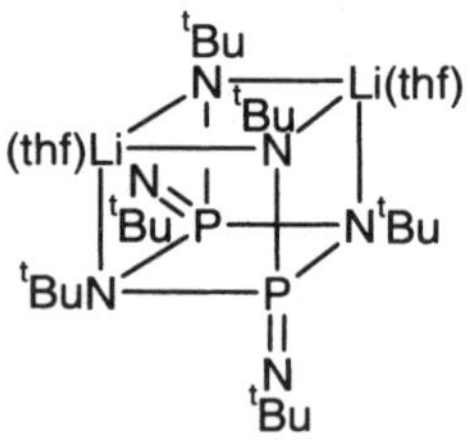

(12)

The potentially versatile ligating ability of trisimidometaphosphate anions is manifested in the formation of either dispirocyclic (**13**) or heterocubane (**14**) rhenium complexes from the reaction of $[(Me_3Si)_2NP(=NSiMe_3)_2]$ with $Re(CO)_5Cl$ [18, 19]. Both (**13**) and (**14**) contain the dianionic $[(RN)_2P(\mu\text{-}NR)_2P(NR)_2]^{2-}$

($R=SiMe_3$) ligand. Apparently, the loss of a CO ligand on each rhenium atom causes the bis(*N,N'*)-chelated complex (**13**) to rearrange to a cube (**14**).

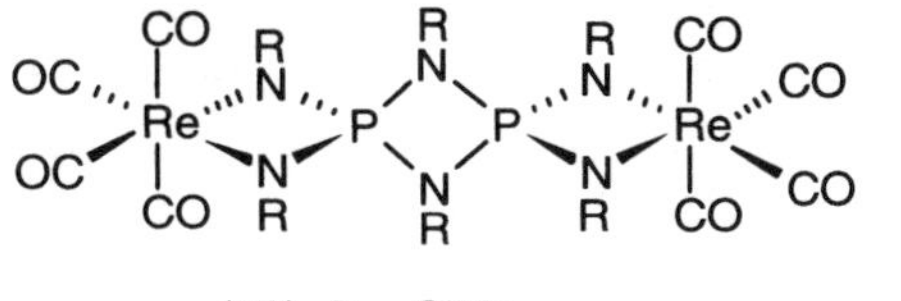

(13) R = $SiMe_3$

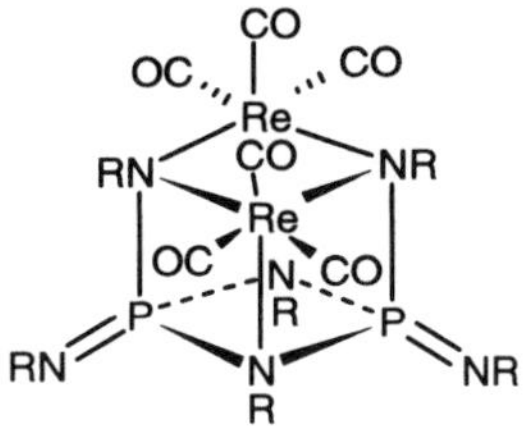

(14) R = $SiMe_3$

Dispirocyclic systems (**15**) are also formed from the reaction of *cis*-[$^tBuNH(ArN)P(\mu$-$N^tBu)_2P(NAr)NH^tBu$] (Ar = Ph, *p*-tolyl) with two equivalents of trimethylaluminum [20].

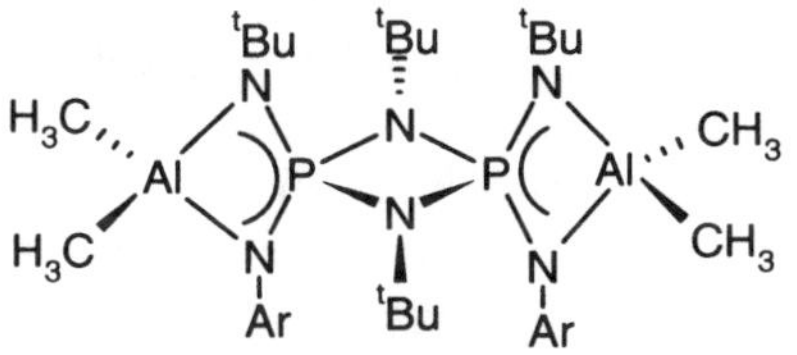

(15) (Ar = C_6H_5, 4-$CH_3C_6H_4$)

2.3
Tetrakis(imido)phosphate Anions $[P(NR)_4]^{3-}$

The first example of a tetrakisimido analogue of the orthophosphate ion, PO_4^{3-}, the solvent-separated ion pair $[(THF)_4Li][(THF)_4Li_2P(Nnaph)_4]$ (**16**), was reported by Russell et al. [21]. This complex was isolated in low yield from the reaction of P_2I_4 with α-naphthylamine in THF/NEt_3, followed by the addition of ^nBuLi. The mechanism of this remarkable redox process is not understood.

A more rational approach to the synthesis of tetrakisimido phosphates involves the trilithiation of $Me_3SiN{=}P(NH^tBu)_3$ (prepared from $Cl_3P{=}NSiMe_2$ and $LiNH^tBu$) with three equivalents of ^nBuLi in THF (Eq. 2) [22]. This method is potentially applicable to a variety of anions of the type $[P(NSiMe_3)(NR)_3]^{3-}$ (R = alkyl, aryl). Although the complex (**17**) was prepared in THF solution, it is noteworthy that the Li^+ ions are not solvated by THF. The structure of (**17**) can be described as a hexagonal prism (or cyclic Li_6N_6 ladder) bicapped on the two hexagonal faces by μ^3 $PNSiMe_3$ groups. Thus the structure bears a striking resemblance to those of the dimeric trisimidopnicogenite clusters $[Li_3\{E(N^tBu)_3\}]_2$ (E = As, Sb) [23, 24]:

$$Me_3SiN{=}P(NH^tBu)_3 + 3\ ^nBuLi \xrightarrow[-3\ ^nBuH]{THF} \{Li_3[P(N^tBu)_3(NSiMe_3)]\}_2 \quad (2)$$

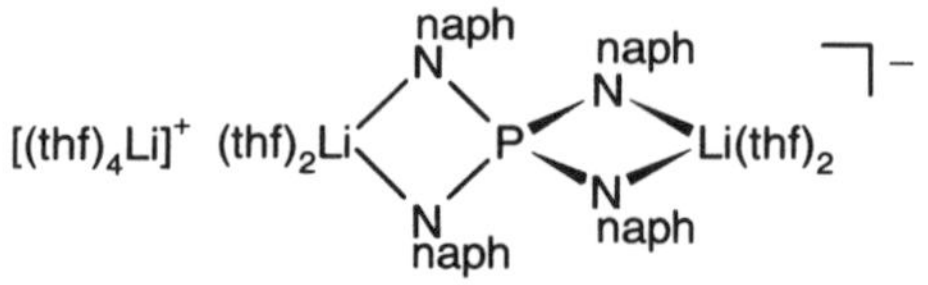

(16)

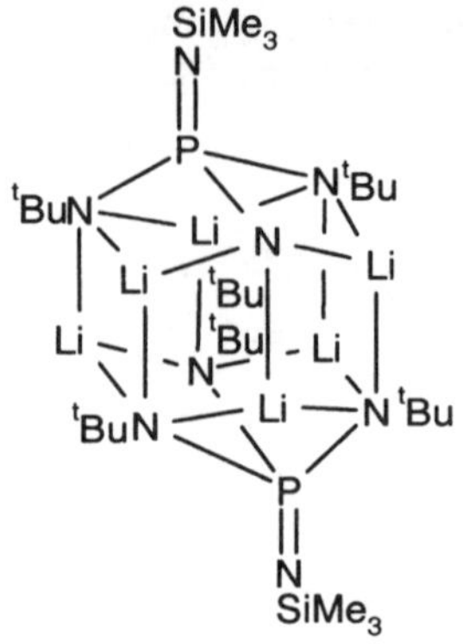

(17)

2.4 Tris(imido)phosphonate Anions $[R'P(NR)_3]^{2-}$

The first example of a trisimido analogue of the phosphite HPO_3^{2-} dianion was obtained recently by Russell et al. from the reaction of PCl_3 with three equivalents of 2-$MeOC_6H_4NH_2$ and subsequent treatment with ^nBuLi [25]. The dimeric complex $\{Li_2[PH(NR)_3]\}_2$ (R=2-methoxyphenyl) (**18**) forms a lantern-like structure in which two tetrahedral $[HP(NR)_3]^{2-}$ anions are linked by four Li^+ ions. Although all four Li^+ ions are tetracoordinate, there are two distinct lithium environments. In one coordination sphere lithium is linked to one imido nitrogen atom from each $[HP(NR)_3]^{2-}$ anion and is additionally complexed by two methoxy sidearms of the aryl substituent. The second coordination mode involves two nitrogens from one $[HP(NR)_3]^{2-}$ anion, one nitrogen from the other $[HP(NR)_3]^{2-}$ anion and one methoxy sidearm. The P–H groups in (**18**) are likely generated via proton shift from an amido group to the phosphorus(III) center [25].

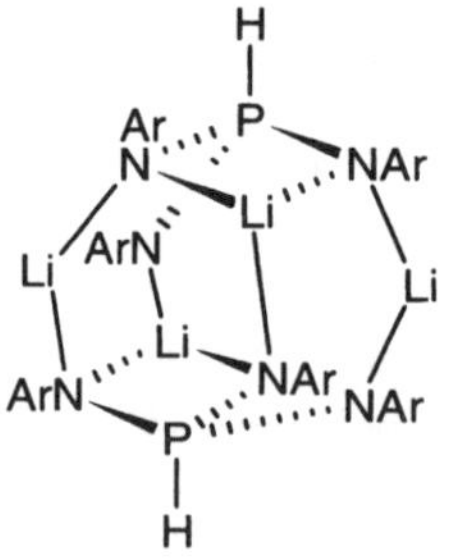

(18) Ar = 2-$CH_3OC_6H_4$

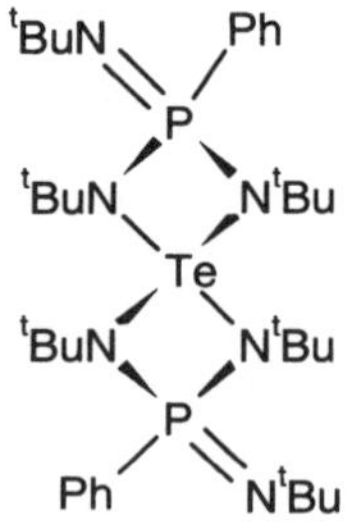

(19)

Although alkali-metal salts of trisimido organophosphates $[R'P(NR)_3]^{2-}$ have not been isolated and structurally characterized compelling evidence for their formation has been presented by Bailey et al. [26]. The reaction of a (trialkylamino)phenylphosphonium salt with three equivalents of ^{n}BuLi in THF produces a yellow solution which, when treated with methyl iodide, yields the trimethylated phosphonium cation $PhP[N(Me)^{i}Pr]_3^+$ (Eq. 3):

$$[PhP(NH^{i}Pr)_3]Br \xrightarrow[\text{(ii) 3 MeI}]{\text{(i) 3 }^{n}\text{BuLi}} [PhP\{N(Me)^{i}Pr\}_3]I + LiBr + 2\,LiI \quad (3)$$

The reaction of the in situ generated reagent $Li_2[PhP(N^{i}Pr)_3]$ with $[(\eta\text{-}p\text{-cymene})RuCl_2]_2$ yielded, after work-up, $[(\eta\text{-}p\text{-cymene})Ru\{\eta^2\text{-}(^{i}PrN)_2PPh(NH^{i}Pr)\}][BPh_4]$ (**26**). Apparently, mono-protonation of the initial reaction product occurred during the isolation procedure.

The only structurally characterized derivative of a trisimido organophosphonate anion is the spirocyclic tellurium(IV) complex (**19**), which is obtained from the interesting redox reaction between $PhPCl_2$ and $[Li_2Te(N^{t}Bu)_3]_2$ [27]. The phosphorus(V)-centered ligands are generated by imide transfer from tellurium to the phosphorus(III) atoms with concomitant reduction of one-half of the tellurium in the Te(IV) reagent to elemental tellurium [27].

2.5 Bis(imido)phosphinite Anions $[R'_2P(NR)_2]^-$

Although the prototypical bisimido analogue of the hypophosphite anion $H_2PO_2^-$ is not known, the organic derivatives $[R'_2P(NR)_2]^-$ (R = alkyl, aryl) have been studied extensively [4]. The bisiminophosphinate anion $[Ph_2P(NSiMe_3)_2]^-$ behaves as a chelating ligand towards tellurium(IV) [28] and in complexes with lanthanides or actinides [29, 30]. Steiner and Stalke have shown that alkali-metal derivatives of the same anion may adopt chelated structures (**20**) or dimeric, stair-shaped arrangements (**21**) in the solid state [31]. In the case of sodium a *bis*-chelated, solvent-separated ion pair $[Na(THF)_6][Na\{(NSiMe_3)_2PPh_2\}_2]$ is formed [31].

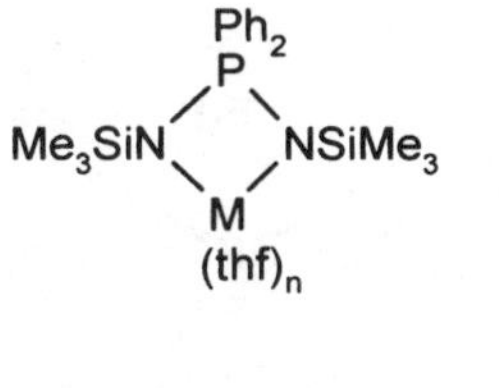

(20) M = Li, n = 2
M = K, n = 4

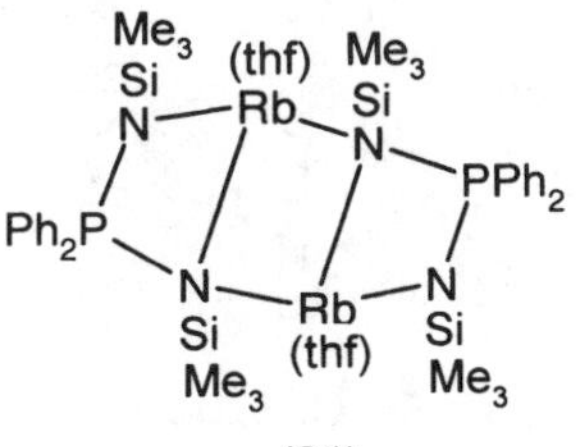

(21)

3 Heteroleptic Imidochalcogenido Anions of Phosphorus

3.1 Imidochalcogenido Analogues of the Metaphosphate Ions PE_3^- (E=O, S, Se)

The replacement of one or two of the chalcogen atoms in the metaphosphate anions PE_3^- (E=O, S, Se) by imido groups generates heteroleptic anions of the type $[PE_2(NR)]^-$ or $[PE(NR)_2]^-$, respectively. There is an extensive chemistry of polymetaphosphimates $(PO_2NH)_n^{n-}$, which has been reviewed recently by Schnick et al. [32]. Like the cyclometaphosphates $(PO_3)_n^{n-}$, the metaphosphimates may exist as six- or eight-membered rings (**22**) and (**23**), respectively, with a variety of conformations; the NH groups form part of the ring system. A more detailed discussion of the structures and metal complexes of polymetaphosphimates is given in [32].

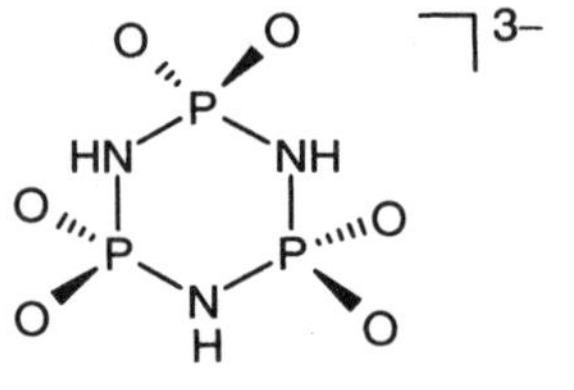

(22)

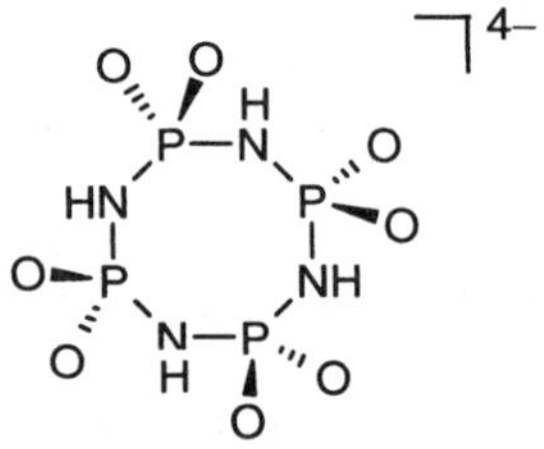

(23)

By contrast to the metaphosphimates, the metaphosphate analogues in which two chalcogens are replaced by imido groups form dimeric, dianionic ligands (**24**). There is only one example of a metal complex of the oxo system (**24**, E=O). An *N,O*-chelated bis(dimethylaluminum) complex is prepared by the reaction of two equivalents of trimethylaluminum with *cis*-[^{t}BuNH(O)P(μ-N^{t}Bu)$_2$P(O)NH^{t}Bu] [20]. The structure of this complex is comparable to that of (**15**).

(24) E = O, S, Se, Te

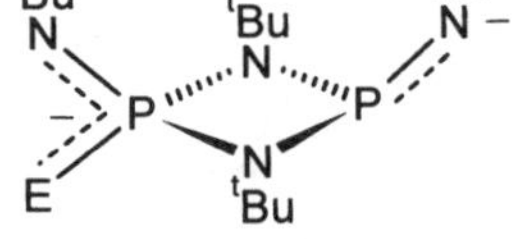

(25) E = Se, Te

The dianions containing heavier chalcogens (**24**, E=S, Se) have been investigated more extensively by Chivers et al. [33]. Alkali-metal derivatives are prepared by the treatment of *cis*-[^{t}BuNH(E)P(μ-N^{t}Bu)$_2$P(E)NH^{t}Bu] with two equivalents of MN(SiMe$_3$)$_2$ (M=Na, K) [34] or ^{n}BuLi[35] (Eq. 4). Bis(dimethylaluminum) complexes are obtained from the corresponding reaction with two equivalents of trimethylaluminum [20]. The use of elevated temperatures in the reaction of *cis*-[^{t}BuNH(Se)P(μ-N^{t}Bu)$_2$P(Se)NH^{t}Bu] with ^{n}BuLi in THF results in deselenation to give (**25**) (E=Se) [35]. This mixed oxidation state [P(III)/P(V)] dianion is better prepared by lithiation of the monoselenide [^{t}BuNH(Se)P-(μ-N^{t}Bu)$_2$P(NH^{t}Bu)] [36]. Interestingly, the mixed oxidation systems (**25**) can be viewed to result formally from the cycloaddition of the hypothetical monomers [P(N^{t}Bu)$_2$]$^-$ (see Sect. 2.1) and [EP(N^{t}Bu)$_2$]$^-$ (E=Se, Te):

cis-[(^{t}BuNH)(E)P(μ-N^{t}Bu)$_2$P(E)NH^{t}Bu] (4)
(E = S, Se)

↓ 2 M[N(SiMe$_3$)$_2$] (M = Na, K), THF

{(THF)K[(^{t}BuN)(E)P(μ-N^{t}Bu)$_2$P(E)(N^{t}Bu)]K(THF)$_x$}
(E = S, x = 1; E = Se, x = 2)

The synthesis of the tellurium analogues of (**24**) and (**25**) requires a different approach, since it is not possible to prepare the necessary amido precursors in significant yields by telluration of [^{t}BuN(H)P(μ-N^{t}Bu)$_2$PN(H)^{t}Bu]. However, the prior lithiation of this P(III)/P(III) system to give (**7**) followed by reaction with elemental tellurium generates the dianion (**24**, E=Te) as its dilithium salt (Eq. 5) [37]:

[Li(THF)]$_2$[^{t}BuNP(μ-N^{t}Bu)$_2$PN^{t}Bu] (5)

↓ Te, TMEDA toluene, 80°

[Li(TMEDA)]$_2$[^{t}BuN(Te)P(μ-N^{t}Bu)P(Te)N^{t}Bu]

The versatile coordinating ability of the ambidentate ligands (**24**) is illustrated nicely by the structures of their alkali-metal complexes. The mode of coordination is determined by the size of the alkali metal cation as well as the nature of the chalcogen. Thus the dilithium derivatives form *bis*(*N,E*)-chelated complexes (**26**) for E=S, Se, whereas the dianionic ligand is *N,N′* and *E,E′* coordinated to the alkali metal in the disodium and dipotassium complexes (**27**) [34, 35]. In the potassium derivatives further aggregation of the building block (**27**) occurs via K-E interactions to give dimers [34]. Intriguingly, the ditelluride ligand (**24**, E=Te) engages in a unique bonding mode in the dilithium complex (**28**). One Li$^+$ cation is coordinated in a *Te,Te′* fashion, while the second Li$^+$ cation is *N, Te*-chelated [37]. A preference for binding to the soft centers of the dianion (**24**, E=S) is displayed by platinum in complex (**29**) [38], whereas the Me$_2$Sn group in (**30**) is coordinated by the hard (*N,N′*) centers [39].

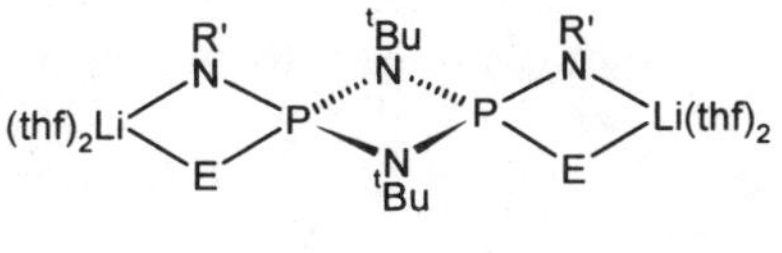

(26)

(27) E = S, Se; M = Na, K

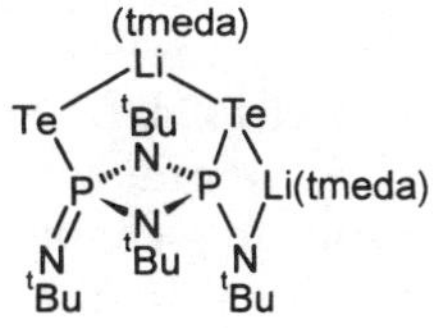

(28)

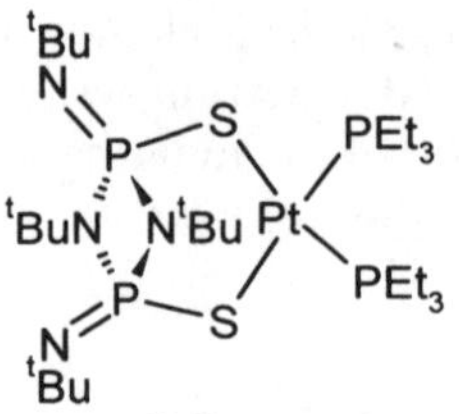

(29)

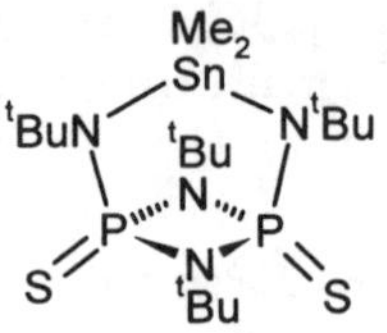

(30)

N,N′-Chelation is also exhibited by the dianionic P(III)/P(V) ligands (**25**) in the Me_2Sn complex (**31**) [39] and in the magnesium complex (**32**) [40], which is prepared by oxidation of {Mg(thf)$_2$[^{t}BuNP(μ-N^{t}Bu)$_2$PN^{t}Bu]} by elemental tellurium [40]. One of the endocyclic N^{t}Bu groups in (**32**) is also weakly coordinated to magnesium, thus providing an intramolecular base-stabilization similar to that observed for complexes of type (**8**).

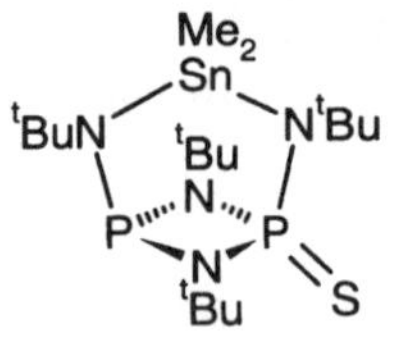

(31)

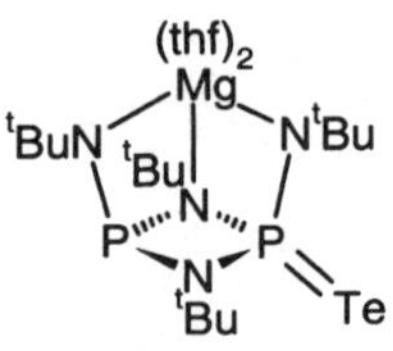

(32)

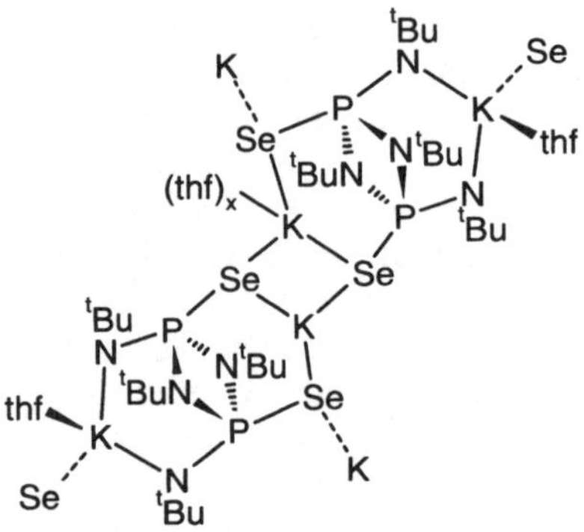

(33)

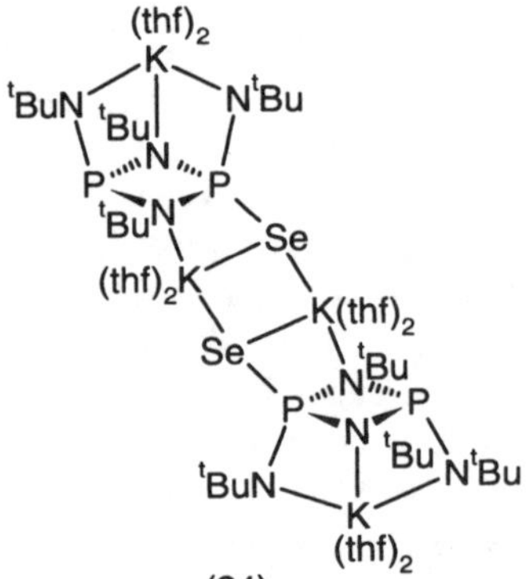

(34)

A comparison of the structures of (**33**) [36] and (**34**) [34] is informative. These complexes are dipotassium salts of the P(V)/P(V) and P(III)/P(V) dianions (**24**, E=Se) and (**25**, E=Se), respectively. Both complexes form dimers with a central K_2Se_2 ring. However, the absence of a second Se donor site in (**34**) results in a change of coordination mode from *Se,Se′* to *μ-N,Se* for the K^+ ions that form the K_2Se_2 ring. In addition, the second K^+ cation is *N,N′,N″* coordinated. Thus, both bridging N^{t}Bu groups of the P_2N_2 ring are involved in bonding to K^+ ions.

3.2
Imidochalcogenido Analogues of Orthophosphate Anions PE_4^{3-} (E=O, S)

The reagents $EP(NHR)_3$ (E=O, S; R=alkyl, aryl) are easily prepared and represent a potential source of the orthophosphate analogues $[EP(NR)_3]^{3-}$. However, in contrast to the behaviour of $(Me_3SiN)P(NH^tBu)_3$ (see Sect. 2.3), complete metallation of the tris(alkylamido) derivatives cannot be achieved with organolithium reagents. The reaction $OP[N(H)^tBu]_3$ with one equivalent of $Li[N(SiMe_3)_2]$ brings about monodeprotonation and the trimer (**35**) is isolated [41]. The ^{31}P NMR spectrum of (**35**) in d_8-THF shows a singlet a δ 5.8 indicating that dissociation occurs in solution to give a more symmetrical species, possibly a centrosymmetric dimer. In the light of this suggestion, it is pertinent to note that treatment of $OP[N(H)^tBu]_3$ with two equivalents of ^nBuLi results in double deprotonation to give the dimeric box-shaped structure (**36**). Double deprotonation also occurs in the reaction of $OP(NH^tBu)_3$ with $LiAlH_4$ in boiling toluene to give the twelve-membered ring (**37**) [42]. The dianionic $[OP(N^tBu)_2(NH^tBu)]^{2-}$ ligands in this heterobimetallic complex adopt two different chelating modes (*N,N* and *N,O*) towards the aluminum center thus forcing the bulky N^tBu groups to the periphery of the structure.

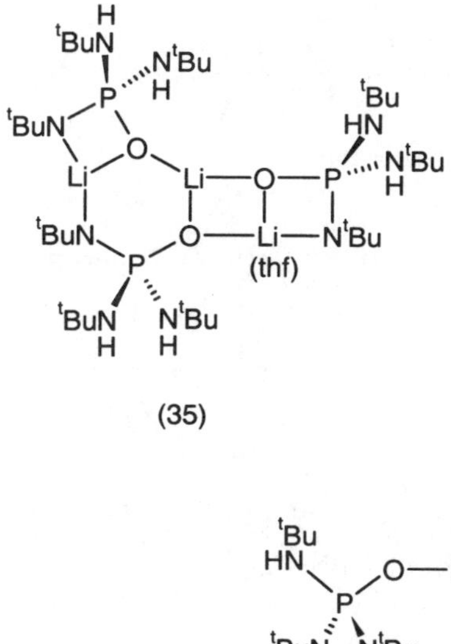

(35)

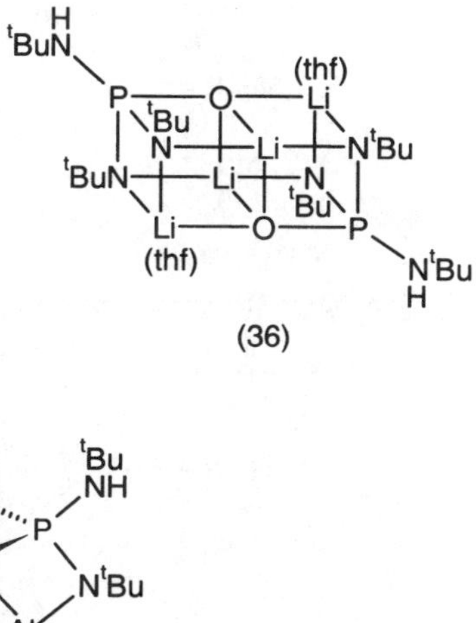

(36)

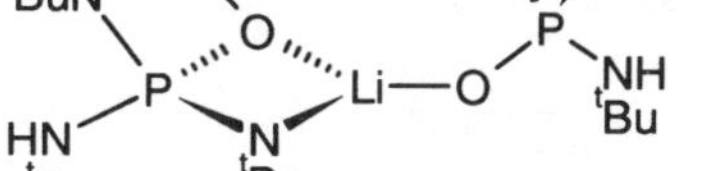

(37)

As indicated in Scheme 1 the outcome of the lithiation of trisamidothiophosphates $SP[N(H)R]_3$ with organolithium reagents is influenced by the steric bulk of the R groups [43]. When R=^{i}Pr both mono- and dilithiated complexes (**38**) and (**39**), respectively, may be obtained by using the stoichiometric amount of ^{n}BuLi. The centrosymmetric dimer (**38**) forms a step-shaped ladder in a transoid conformation via Li-N interactions. Significantly, chelation of Li^+ ions by TMEDA ligands in (**39**) prevents laddering via Li-N interactions. The polycyclic cluster (**40**) contains both mono- and di-anions $[SP(N^tBu)_x(NH^tBu)_3]^{x-}$ (x=1, 2). By contrast to the dimer (**38**), which aggregates via Li-N interactions, the central feature of (**40**) is a central cisoid Li_4S_4 ladder. The lability of the P=S bond in trisamidothiophosphates is illustrated by the formation of the cubane (**12**) upon reaction of SP[N(H)^{t}Bu] with two equivalents of ^{n}BuLi [43]. The trianion $[SP(NR)_3]^{3-}$ may however be generated by changing the R substituents to aryl groups, e.g. by trilithiation of SP[N(H)*p*-tolyl] [41]. The electron-withdrawing aryl substituents enhance the acidity of the NH_3 protons.

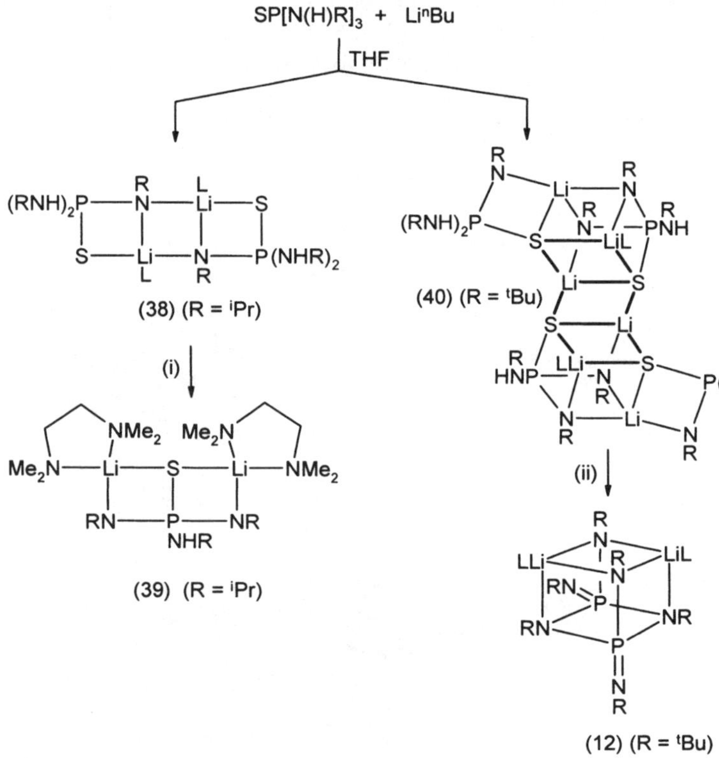

Scheme 1

3.3
Imidochalcogenido Analogues of Phosphonate Anions $[RPE_3]^{2-}$ (E=O, S, Se, Te)

Organophosphonate anions RPO_3^{2-} are important ligands in the formation of layered structures [44]. Although little is known about metal complexes of the trisimido analogues $[RP(NR)_3]^{2-}$ (see Sect. 2.4), Chivers et al. have reported that heteroleptic dianions of the type $[PhP(E)(N^tBu)_2]^{2-}$ (**40**, E=O, S, Se) are readily generated by the treatment of $PhP(E)[N(H)^tBu]_2$ with two equivalents of ^nBuLi in THF [45]. The tellurium analogue (**41**, E=Te) is conveniently prepared by the reaction of $\{Li_2[PhP(N^tBu)_2]\}_2$ [46] with elemental tellurium:

$$PhP(E)(NH^tBu)_2 + 2\ ^nBuLi \xrightarrow{THF} [Li(THF)]_2[PhP(E)(N^tBu)_2] \quad (E = S, Se) \tag{6}$$

$$Li_2[PhP(N^tBu)_2] + Te \xrightarrow{THF} [Li(THF)]_2[PhP(Te)(N^tBu)_2] \tag{7}$$

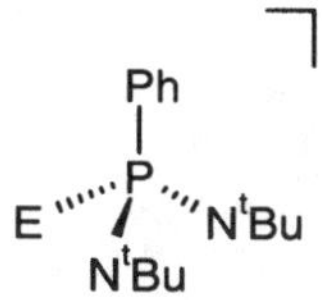

(41) E = O, S, Se, Te

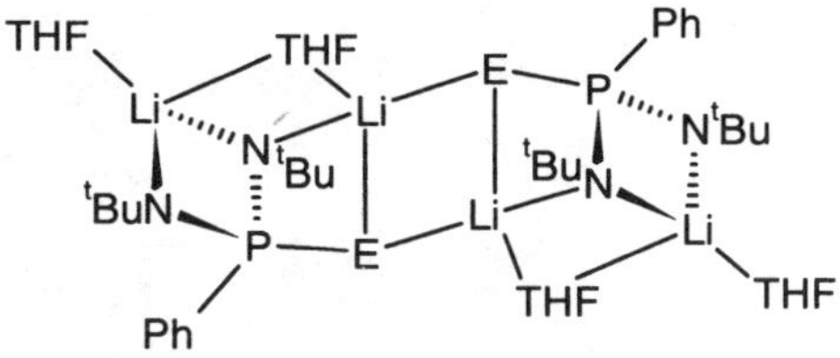

(42) E = Se, Te

The dilithium derivatives (**42**, E=S, Se, Te) all form dimers that are linked through Li-E interactions. The selenide and telluride (**42**, E=Se, Te) contain central transoid Li_2E_2 rings, whereas the sulfide (**42**, E=S) has only very weak, transannular S...S interactions [45]. By contrast, the dilithium derivative of the oxide (**41**, E=O) is obtained as an Li_2O-templated tetramer with a μ_6-OLi_6^{4+} core (**43**) (see Fig. 1). This structural feature has been observed in a variety of lithium-nitrogen clusters, e.g. $Li_4[^nBuC(N^tBu)_2]_4{\cdot}Li_2O$ [47], in which the source of Li_2O has been shown to be adventitious water. The structure of (**43**) provides a nice illustration of the potentially versatile ligating ability of the phosphonate analogue (**40**, E=O). The four phosphonate anions in (**43**) are on the periphery of the cluster and display two different coordination modes. In each mode the phosphonate ligands coordinate five different Li^+ ions through the two imido nitrogens and the single oxygen atom. One pair of phosphonate ligands is *N,N′*-chelated to one Li^+ ion, while the other pair is *N,O*-chelated to a Li^+ ion; further coordination occurs through both the oxygen and nitrogen atoms to four other Li^+ ions. When the solvent used for the lithiation of $PhP(O)[N(H)^tBu]_2$ is changed from THF to diethyl ether the tetrameric cluster (**44**) is obtained (see Fig. 2) [45]. The low symmetry of (**44**) is indicated by the variety of coordination numbers observed for Li (3 and 4), O (3 and 5) and N (3, 4 and 5) in the monosolvated structure.

Analogous to the formation of (**15**), the reaction of $PhP(E)[N(H)^tBu]_2$ with one equivalent of trimethylaluminum produces the complexes (**45**, E=S, Se) in which the ligands (**41**, E=S, Se) are *N,E*-chelated to aluminum [20].

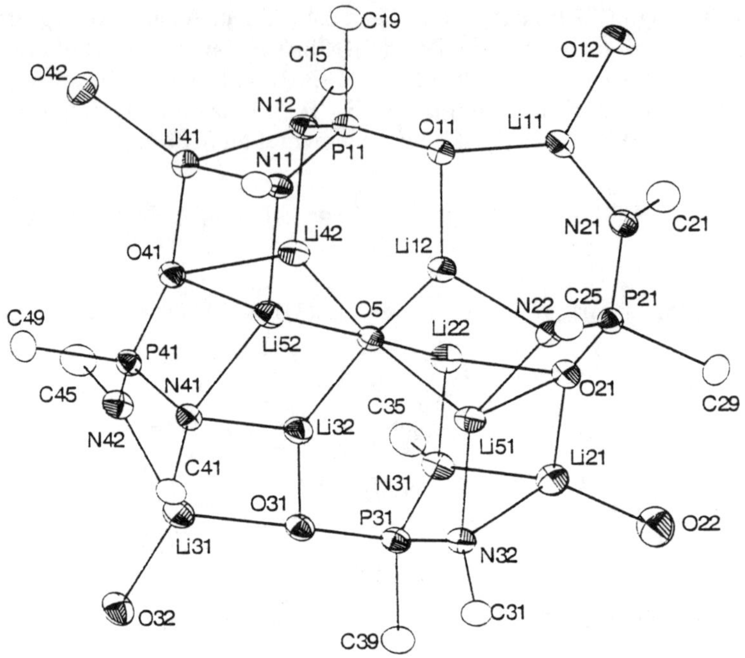

Fig. 1. X-ray structure of $\{(THF)Li_2[PhP(O)(N^tBu)_2]\}_4 \cdot Li_2O$ (**43**). Only the α-carbon atoms of the ^{t}Bu groups, the ipso-carbons of phenyl groups and the oxygen atoms of THF molecules are shown

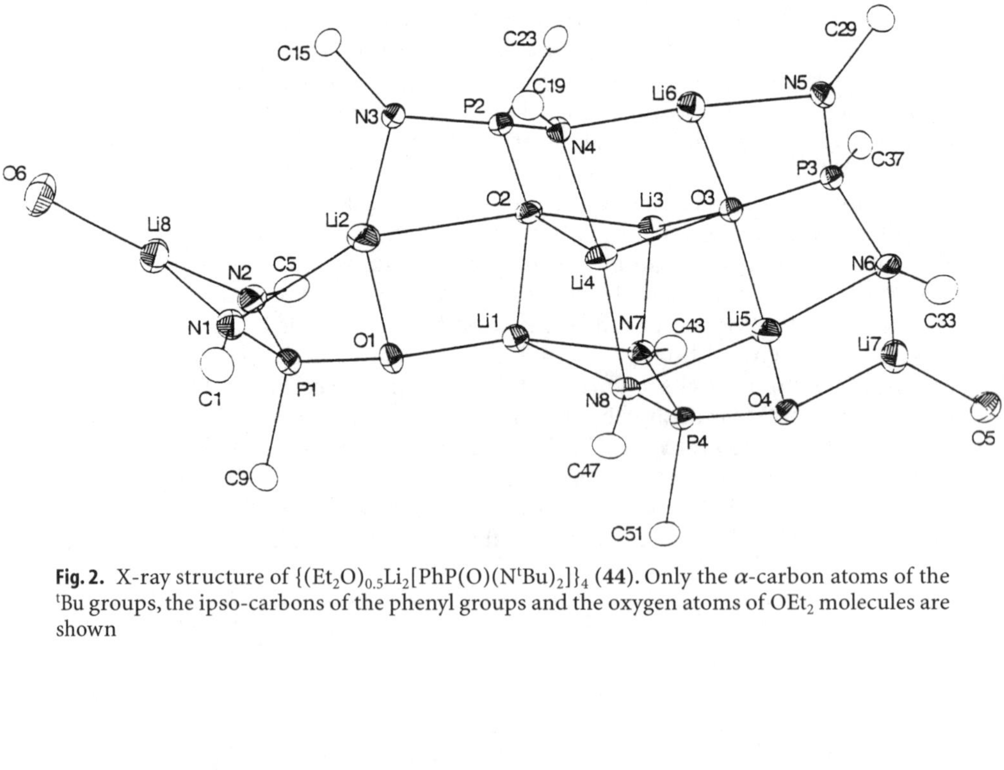

Fig. 2. X-ray structure of $\{(Et_2O)_{0.5}Li_2[PhP(O)(N^tBu)_2]\}_4$ (**44**). Only the α-carbon atoms of the ^{t}Bu groups, the ipso-carbons of the phenyl groups and the oxygen atoms of OEt_2 molecules are shown

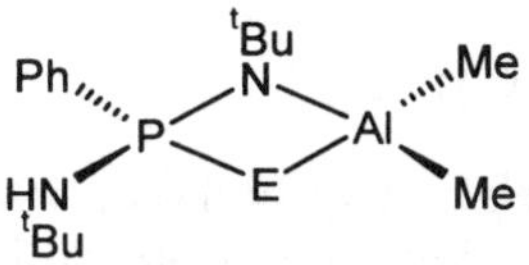

(45) E = S, Se

3.4 Imidochalcogenido Analogues of the Phosphinate Anions $[R_2PE_2]^-$ (E=O, S, Se, Te)

The coordination chemistry of dithiophosphinates ($R_2PS_2^-$) has attracted considerable attention because of the ability of these anionic ligands to chelate and/or bridge metal centers. Only a few investigations of the hybrid systems (**46**), which combine hard (N) and soft (E=S, Se, Te) centers, have been reported. The anion (**46**, E=S, R=$SiMe_3$) has been structurally characterized as the Ph_4As^+ salt, which was prepared by the reaction of $Ph_2P(S)N(SiMe_3)_2$ with one equivalent of KO^tBu in THF followed by metathesis with $[AsPh_4]Cl$ (Eq. 8) [48]. The potassium derivative of the selenium analogue (**47**) has been synthesized in a similar manner and shown to have a dimeric structure with a central K_2Se_2 ring [49]. Anions of the type (**46**, E=S) behave as chelating ligands in the octahedral tin(IV)complex $[(^iPr_2P(S)NSiMe_3)_2SnCl_2]$ [50] and in square planar to tetrahedral nickel(II) and cobalt(II) complexes [51]:

$$Ph_2P(S)N(SiMe_3)_2 + KO^tBu \xrightarrow[THF]{[Ph_4As]Cl} [Ph_4As][Ph_2P(S)NSiMe_3] + Me_3SiO^tBu + KCl \quad (8)$$

$$^tBu_2P(Te)NHR' + {^nBuLi} \longrightarrow Li[^tBu_2P(Te)NR'] + {^nBuH} \quad (9)$$

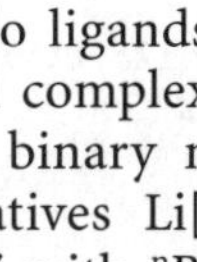

(46) E = S, Se, Te

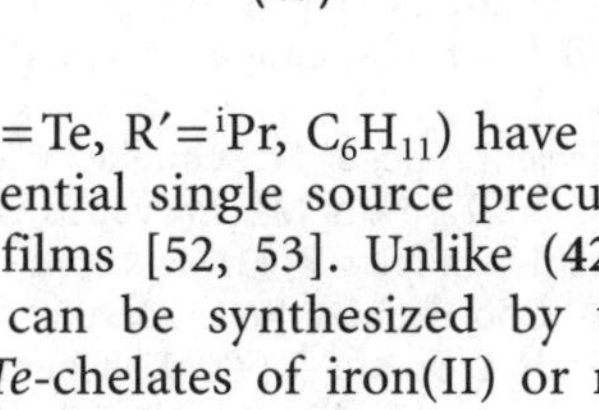

(47)

The tellurido ligands (**46**, R=^tBu, E=Te, R′=^iPr, C_6H_{11}) have been investigated in metal complexes that are potential single source precursors for the generation of binary metal telluride films [52, 53]. Unlike (**42**, E=Te) the lithium derivatives $Li[^tBu_2P(Te)NR']$ can be synthesized by treatment of $^tBu_2P(Te)NHR'$ with ^nBuLi (Eq. 9). *N,Te*-chelates of iron(II) or nickel(II) are obtained by metathesis with $MCl_2(PMe_3)_2$ (M=Fe, Ni). Alternatively, Cr, Mn, Co, Zn or Cd complexes can be prepared by the direct reaction of $^tBu_2P(Te)NHR'$ with $M[N(SiMe_3)_2]$.

4 Conclusions and Future Directions

The development of the chemistry of homoleptic polyimido and heteroleptic imido-oxo or -chalcogenido anions of phosphorus is at an early stage. The availability of a convenient NMR probe, i.e. ^{31}P, is a major advantage in the study of phosphorus-centered polyimido anions compared to those based on other p-block elements. The metallation of appropriate amido precursors by reagents such as $^{n}BuLi$ or $MN(SiMe_3)_2$ provides a facile, high yield method of generating isoelectronic analogues of common, as well as metastable, oxo anions of phosphorus, e.g. PO_2^-, PO_3^-, PO_4^{3-}, RPO_3^{2-} and $R_2PO_2^-$. The structural information available to date is primarily restricted to alkali-metal derivatives. Nevertheless, the versatile coordination behaviour of these multidentate ligands is already apparent, especially for the heteroleptic imidochalcogenido systems. Complexes of these novel anions with p- or d-block metals have received little attention. Topics that merit investigation include (a) the influence of these novel anions on the catalytic behaviour of transition metal centers, (b) the stabilization of unusual geometries in heavy p-block element complexes, and (c) the incorporation of dianions, such as (**24**) or (**25**), into metal-containing polymers with extended structures. The heteroleptic systems studied to date are those with a high RN:E ratio (E=chalcogenido). Important target species include anions with a low RN:E ratio, e.g. $[P(NR)O_3]^{3-}$, in order to investigate the influence of the replacement of an oxo ligand by an NR substituent on the structure and properties of porous materials like aluminophosphates. Finally, it is noted that some of these cluster molecules, e.g. (**17**) and (**36**) [22], form intensely coloured solutions upon exposure to oxygen suggested the formation of radical systems. This observation is reminiscent of the behaviour of the analogous chalcogen-containing systems $\{Li_2[E(N^tBu)_3]\}_2$ (E=S, Se) [54, 55]. The identification of these phosphorus-containing radicals is a worthwhile challenge.

5 References

1. Greenwood NN, Earnshaw A (1997) Chemistry of the elements, 2nd edn. Butterworth Heineman, Oxford
2. Brask JK, Chivers T (2001) Angew Chem Int Ed Engl 40:3960
3. Aspinall GM, Copsey MC, Leedham AP, Russell CA (2002) Coord Chem Rev 227:217
4. Steiner A, Zacchini S, Richards PI (2002) Coord Chem Rev 227:193
5. Hitchcock PB, Jasim HA, Lappert MF, Williams HD (1986) J Chem Soc Chem Commun 1634
6. Detsch R, Niecke E, Nieger M, Schoeller WW (1992) Chem Ber 125:1119
7. Brask JK, Chivers T, Krahn ML, Parvez M (1999) Inorg Chem 38:290
8. Schranz I, Stahl L, Staples RJ (1998) Inorg Chem 37:1493
9. Stahl L (2000) Coord Chem Rev 210:203
10. Grocholl L, Stahl L, Staples RJ (1997) Chem Commun 1465
11. Grocholl L, Schranz I, Stahl L, Staples RJ (1998) Inorg Chem 37:2496
12. Moser DF, Schranz I, Gerrety MC, Stahl L, Staples RJ (1999) J Chem Soc Dalton Trans 751
13. Schranz I, Moser DF, Stahl L, Staples RJ (1999) Inorg Chem 38:5814
14. Moser DF, Carrow CJ, Stahl L, Staples RJ (2001) J Chem Soc Dalton Trans 1246

15. Lief GR, Carrow CJ, Stahl L, Staples RJ (2001) Chem Commun 1562
16. Niecke E, Frost M, Nieger M, von der Gönna V, Ruban A, Schoeller WW (1994) Angew Chem Int Ed Engl 33:2111
17. Chivers T, Krahn M, Parvez M, Schatte G (2001) Chem Commun 1922
18. Scherer OJ, Quintus P, Sheldrick WS (1987) Chem Ber 120:1183
19. Scherer OJ, Quintus P, Kaub J, Sheldrick WS (1987) Chem Ber 120:1463
20. Lief GR, Carrow CJ, Stahl L, Staples RJ (2001) Organometallics 20:1629
21. Raithby PR, Russell CA, Steiner A, Wright DS (1994) Angew Chem Int Ed Engl 36:649
22. Armstrong A, Chivers T, Krahn M, Parvez M (2002) Chem Commun 2332
23. Beswick MA, Kidd SJ, Paver MA, Raithby PR, Steiner A, Wright DS (1999) Inorg Chem Commun 2:162
24. Beswick MA, Choi N, Harmer CN, Hopkins AD, McPartlin M, Paver MA, Raithby PR, Steiner A, Tombul M, Wright DS (1998) Inorg Chem 37:2177
25. Burke LT, Hevia-Frere E, Holland R, Jeffery JC, Leedham AP, Russell CA, Steiner A, Zagorski A (2000) Chem Commun 1769
26. Bailey PJ, Grant KJ, Parsons S (1998) Organometallics 17:551
27. Chivers T, Gao X, Parvez M (1995) Angew Chem Int Ed Engl 34:2549
28. Chivers T, Doxsee DD, Gao X, Parvez M (1994) Inorg Chem 33:5678
29. Recknagel A, Witt M, Edelmann FT (1989) J Organomet Chem 371:C40
30. Recknagel A, Steiner A, Noltemeyer M, Brooker S, Stalke D, Edelmann FT (1991) J Organomet Chem 414:327
31. Steiner A, Stalke D (1993) Inorg Chem 32:1977
32. Marchand R, Schnick W, Stock N (2000) Adv Inorg Chem 50:193
33. Briand GG, Chivers T, Krahn M (2002) Coord Chem Rev 233/234:232
34. Chivers T, Krahn M, Parvez M (2000) Chem Commun 463
35. Chivers T, Krahn M, Parvez M, Schatte G (2001) Inorg Chem 40:2547
36. Chivers T, Krahn M, Schatte G (2002) Inorg Chem 41:4348
37. Briand GG, Chivers T, Parvez M (2002) Angew Chem Int Ed Engl 41:2468
38. Chivers T, Fedorchuk C, Krahn M, Parvez M, Schatte G (2001) Inorg Chem 40:1936
39. Linti G, Nöth H, Schneider E, Storch W (1993) Chem Ber 126:611
40. Briand GG, Chivers T, Parvez M, Schatte G (2003) Inorg Chem 42:525
41. Chivers T, Krahn M, Parvez M (2003) Inorg Chem 42: Advance Article, May 25
42. Blais P, Chivers T, Schatte G, Krahn M (2002) J Organomet Chem 646:107
43. Chivers T, Krahn M, Parvez M, Schatte G (2001) Chem Commun 1992
44. Alberti G (1996) Layered metal phosphonates and covalently pillared diphosphonates. In: Comprehensive supramolecular chemistry, vol 7. Pergamon, New York, p 151
45. Briand GG, Chivers T, Krahn M, Parvez M (2002) Inorg Chem 41:6808
46. Eichhorn B, Nöth H, Seifert T (1999) Eur J Inorg Chem 2355
47. Chivers T, Downard A, Yap GPA (1998) J Chem Soc Dalton Trans 2603
48. Chivers T, Parvez M, Seay MA (1995) Z Anorg Allg Chem 621:1813
49. Chivers T, Parvez M, Seay MA (1994) Inorg Chem 33:2147
50. Raubold T, Freitag S, Herbst-Irmer R, Roesky HW (1993) Z Anorg Allg Chem 619:951
51. Deeg A, Kuchen W, Langsch D, Mootz D, Peters W, Wunderlich H (1991) Z Anorg Allg Chem 606:919
52. Bochmann M, Bwembya GC, Whilton N, Song X, Hursthouse MB, Coles SJ, Karanlov AC (1995) J Chem Soc Dalton Trans 1887
53. Bochmann M, Bwembya GC, Hursthouse MB, Coles SJ (1995) J Chem Soc Dalton Trans 2813
54. Fleischer R, Freitag S, Stalke D (1998) J Chem Soc Dalton Trans 193
55. Brask JK, Chivers T, McGarvey B, Schatte G, Sung R, Boeré RT (1998) Inorg Chem 37:4633

Top Curr Chem (2003) 229: 161–198
DOI 10.1007/b11155

Recent Advances in the Chemistry of Difunctionalized Organo-Phosphorus and -Sulfur Compounds

Mihaela Gulea · Serge Masson

Laboratoire de Chimie Moléculaire et Thio-organique (UMR CNRS 6507), ISMRA-Université de Caen, 6 Boulevard du Maréchal Juin, F-14050 Caen, France
E-mail: mihaela.gulea@ismra.fr

Recent results related to the synthesis and reactivity of organocompounds substituted by both phosphorus and sulfur functional groups are reviewed. Main functions associated in these molecules are dithioester, thioamide, thiol, sulfide or sulfoxide and phosphonate, phosphonamide, phosphine or phosphine-oxide. Phosphonodithioformates and dithioacetates are shown to be versatile building blocks through their use in various reactions such as thiophilic addition of nucleophiles, thioacylation of amines and hetero-Diels-Alder reactions. Among other applications, these dithioesters can be used as precursors of new phosphorylated thia-substituted heterocycles and phosphonate analogues of glycoside or nucleoside monophosphates. [1,2]-, [1,3]-, or [2,3]-sigmatropic rearrangements, radical reactions, and asymmetric synthesis involving mixed P and S-organophosphorus compounds are reviewed. The generation of phosphonomethyl and phosphonodifluoromethyl radicals (by cleavage of a phosphorylated carbon-sulfur bond) and their addition to olefinic compounds are reported. The various ways to generate α- or β-thio substituted phosphorylated compounds, in particular enantioselective processes, are listed together with recent examples of the use of chiral phosphorus and sulfur bidentate ligands in asymmetric catalysis.

Keywords. Phosphorus and sulfur derivatives, Dithioesters, Hetero Diels-Alder, Sigmatropic rearrangements, Chiral ligands

1 Introduction

Organophosphorus and organosulfur compounds occupy a broad area in organic chemistry: reagents for well-known reactions or useful synthetic intermediates (phosphines, phosphonates, sulfoxides, sulfones, sulfonyl chlorides, thiocarbonyl compounds, etc.), bioactive molecules or organic materials, which have been the subject of numerous books and monographs. As far as mixed phosphorus and sulfur molecules are concerned, they are well known as reagents for sulfurization (P_4S_{10}, Lawesson and Davy's reagents), pesticides (thio- and dithiophosphate derivatives: malathion, fenthion, carbophenothion, demeton) and also, more recently, as interesting bio-active molecules. Some examples of the latter are presented in Scheme 1: **I** (tiludronate), anti-inflammatory and anti-rheumatism agent [1]; **II**, inhibitor of threonine synthase [2]; **III**, inhibitor of endopeptidase

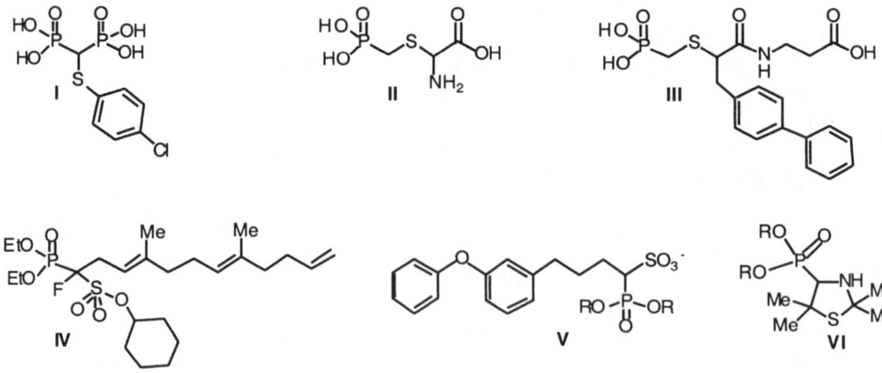

Scheme 1

24.11 [3]; **IV** and **V**, inhibitors of cholesterol biosynthesis [4]; **VI**, anti-inflammatory agents [5].

In the recent literature, an increased interest for mixed phosphorus and sulfur substituted synthetic tools or building blocks has emerged, and also new interesting properties and applications have been recently developed from these compounds. Previous overviews dealing with the chemistry of phosphonodithioformates [6a] and with the reactions of phosphites with thiocarbonyl derivatives [6b] have been written by one of us. Beside, reviews related to α-phosphoryl-substituted organosulfur compounds [7a], α-phosphonate carbanions [7b], and some applications of phosphorus and sulfur compounds in organic synthesis [7c] have been published by Mikolajczyk and Balczewski. As a complement, our article summarizes the more recent results (last ten years) related to difunctionalized organo-phosphorus- and -sulfur compounds in which each function has an influence on the other (this effect inducing specific reactivities) or when properties of both functions are usefully involved in synthetic applications. This, of course, implies compounds in which the two functions are close together in the molecule, i.e., more often, when they are substituents of carbons in alpha or beta position. A phosphonate, phosphonamide, phosphine-oxide, or phosphine function together with a thiol, sulfide, sulfoxide, dithioester, or thioamide group are frequently encountered in these compounds. Our review will focus on four main subjects which are: the rich *chemistry of phosphonylated dithioesters* and three type of reactions which have been the subject of recent studies: *carbanionic sigmatropic rearrangements*, *radical reactions*, and *asymmetric syntheses*, all involving mixed organo-phosphorus and -sulfur compounds.

2 Phosphonodithioesters

Beside thioamides, dithioesters are the most stable and accessible thiocarbonyl compounds. Their specific reactivity, in particular towards nucleophilic reagents and their applications to the formation of carbon-carbon bonds, have already been reviewed [8]. However, as shown below, the presence of a phosphonate function alpha or beta to the thiocarbonyl group in phosphonodithioformates and phosphonodithioacetates makes these difunctional compounds very versatile building blocks. Moreover, for the phosphonodithioacetates, the substitution of the methylenic hydrogen atoms by fluorine increases again their potential as intermediates for the synthesis of modified natural and bioactive phosphorylated structures.

2.1 Phosphonodithioformates

Alkyl (O,O′-dialkylphosphono)methanedithioates or phosphonodithioformates **1** (R^1=Et or *i*Pr) are stable and easily accessible compounds from dialkylphosphites and carbon disulfide [9]. Most of the synthetic applications of these difunctional compounds result from their reactivity towards nucleophiles and dienes.

2.1.1
Reaction with Nucleophilic Reagents

In these α-phosphorylated dithioesters, the electron-withdrawing effect of the phosphono group, which strongly increases the electrophilic character of the thiocarbonyl group, makes the latter more prone to the thiophilic attack of nucleophiles and stabilizes the resulting carbanion. The main reactions of **1** with nucleophiles are summarized in Scheme 2.

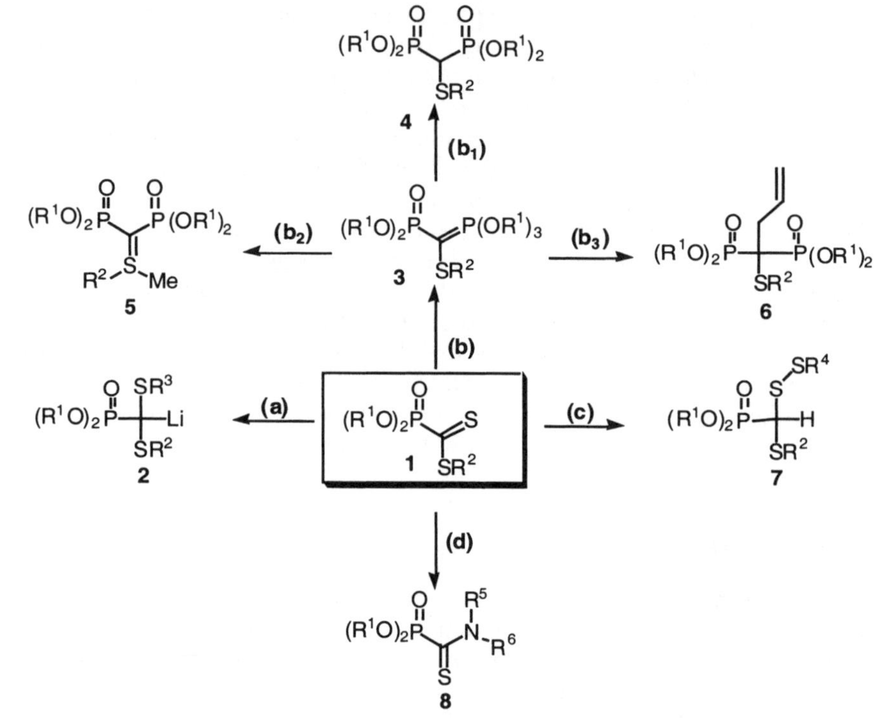

(a) : R^3Li; **(b)**: $2(R^1O)_3P$; **(b$_1$)**: HCl; **(b$_2$)**: MeI; **(b$_3$)**: allylBr; **(c)**: R^4SH, ε NEt_3; **(d)** R^5R^6NH

Scheme 2

The thiophilic addition of organolithium compounds (**a**) to the thiocarbonyl group of **1** leads to lithiated phosphonodithioacetals **2**, which are Horner-Wadsworth-Emmons reagents for the synthesis of ketene dithioacetals [10]. Treated by two equivalents of a trialkylphosphite (**b**), dithioformate **1** is converted, through desulfurization, to a stable phosphonium ylide **3**, which is a precursor of various methylenediphosphonate derivatives. For example, the reaction of the ylide **3** with HCl (**b$_1$**) leads to a C-protonation, together with an Arbuzov type dealkylation of the phosphonium moiety, giving alkylthiomethylene diphosphonates **4**. On the other hand, by addition of MeI (**b$_2$**), an S-methylation and the Arbuzov type dealkylation gives quantitatively the highly stabilized sulfonium ylides **5**. When iodomethane is replaced by an allylic bromide ($\mathbf{b_3}$), the S-allylation of **3** is followed by a spontaneous [2,3]-sigmatropic rearrangement and finally C-allylic alkylthiomethylene diphosphonates **6** have been obtained. Allenic analogues can also be prepared by reaction of **3** with a propargylic halide [11]. Again, due to the electron-withdrawing effect of the phosphono group, thiophilic addition of thiols (**c**), catalyzed by triethylamine, occurs readily (this thiophilic addition is not observed with non-activated dithioesters) to give stable phosphonodithioacetal-disulfides **7**. However, with amines (**d**), the usual C-addition and elimination of the alkythio group is observed leading to phosphonothioamides (or thiocarbamoylphosphonates) **8**.

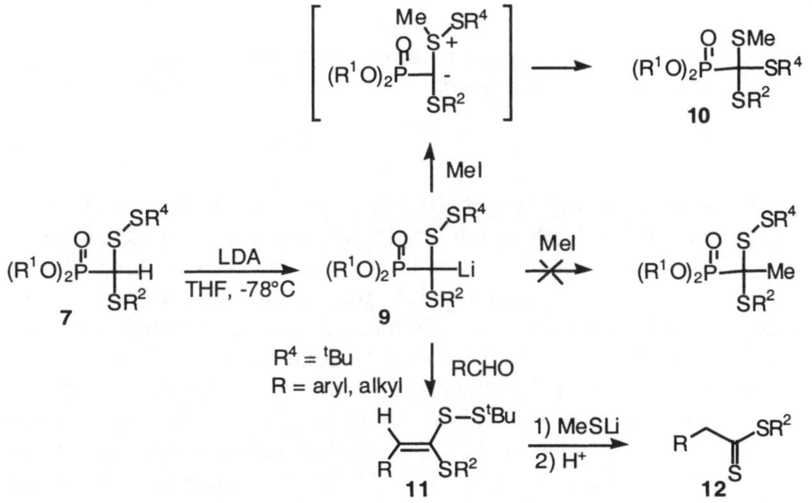

Scheme 3

Several applications of reactions listed in Scheme 2 have been developed. For example, it has been shown that the phosphonodithioacetal-disulfides **7** can be lithiated (Scheme 3). Although, due to the reversibility of the addition of alkylthiolates, the carbanion formed **9** is in equilibrium (strongly favoring **9**) with the starting dithioformate and the lithium alkylthiolate, it can be alkylated or added to a carbonyl compound. Amazingly from carbanion **9**, no alkylation on carbon is observed but only the formation of (trialkylthiomethyl)phosphonate **10**. This is explained by the alkylation of one of the sulfur atoms of the disulfide function followed by a [1,2]-Stevens type rearrangement of the intermediate sulfonium ylide. Treatment of carbanion **9** (R^4=*tert*-butyl) by an aldehyde leads to a ketene dithioacetal-disulfide **11**, the latter being easily cleaved by lithium methylthiolate to give, after protonation, dithioesters **12**. As dithioesters can be easily transformed into dithioacetals by thiophilic addition of Grignard reagents, this reaction is the equivalent of one-carbon homologation with respect to the starting aldehyde [12].

The thioacylating properties of phosphonodithioformate **1** were applied to a variety of functionalized amines as shown in Scheme 4. Phosphonothioacylated

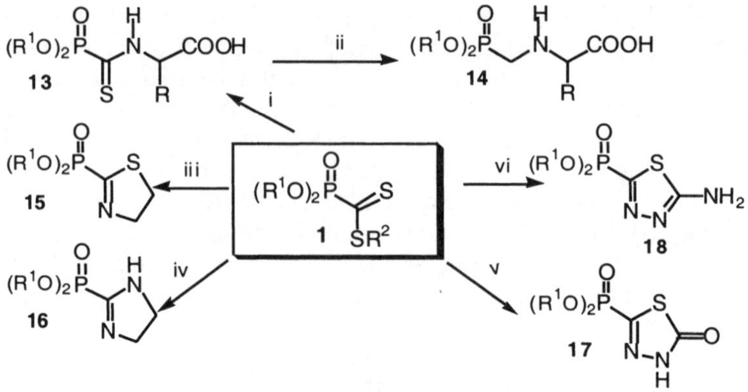

i) aminoacid, NEt_3; ii) $NiCl_2/NaBH_4$; iii) $NH_2CH_2CH_2Br$; iv) $NH_2CH_2CH_2NH_2$
v) $NH_2NHCONH_2$; vi) $NH_2NHCSNH_2$

Scheme 4

aminoacids **13** can be readily prepared. Desulfurization of **13** by $NiCl_2/NaBH_4$ allows the preparation of *N*-phosphonomethylglycine (glyphosphate, Round-up) and other aminoacid derivatives of type **14** [13]. When the thioacylation of amines, functionalized with an appropriate function, is followed by spontaneous cyclization and elimination, new phosphonylated heterocycles such as thiazoline **15**, imidazoline **16**, thiadiazolone **17**, amino thiadiazole **18** are obtained [14].

The sodium salt of sulfanylmethylphosphonate **19** can be easily prepared by reduction of phosphonodithioformate **1** with $NaBH_4$ in refluxing acetonitrile (Scheme 5); the reduction can be halted at the hemidithioacetal stage if the reaction is performed at room temperature [15]. From **19**, various derivatives such as thiol **20**, trithiocarbonates **21**, allylic sulfides **22**, or phosphonomethyl phosphorothiolate **23** can be obtained. Compounds **22** and **23** are starting materials for C–C and C–P bond formation via carbanionic sigmatropic rearrangements (see Sect. 3). The salt **19** can also be obtained directly by reduction with $NaBH_4$

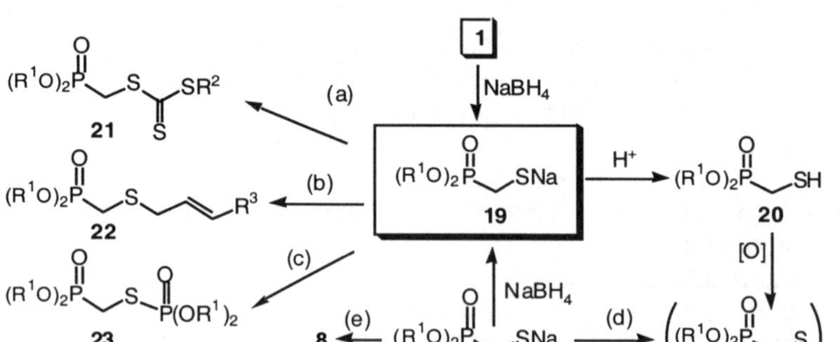

(a) i: CS_2, ii: R^2X; (b) allylic bromide; (c) dialkylchlorophosphate; (d) BCl_3 / AlkSH; (e) BCl_3 / NHR^5R^6

Scheme 5

of the preformed sodium phosphonodithioformate. We should mention that the latter gives, with amines in the presence of BCl_3 as catalyst, the expected phosphonothioamides **8**. Surprisingly, from the reaction of the same sodium phosphonodithioformate with a thiol and the same catalyst, the disulfide of **20** has been obtained [16]. This reaction does not occur with a dithioester non-activated by an electron-withdrawing group and again a mechanism involving thiophilic addition of the thiol to the thiocarbonyl group is proposed.

The reactivity of phosphonodithioformates towards radicals is reported in Sect. 4.

2.1.2
Phosphonodithioformates as Heterodienophiles

The dienophilic properties of thiocarbonyl compounds have been recently reviewed and it is known than dithioesters are less reactive than thioaldehydes or thioketones in [4+2] cycloadditions [8c]. However, due to the electron-withdrawing effect, which lowers the LUMO of the thiocarbonyl group, phosphonodithioformate **1** ($R^1 = i$Pr) is a good heterodienophile (Scheme 6). It reacts with butadiene, isoprene, or dimethylbutadiene at room temperature to give phosphonylated and sulfanylated dihydrothiopyrans **24** in excellent yields [17]. With cyclopentadiene a 70/30 mixture of bicyclic adducts, in favor of the exo-phosphono derivative, has been isolated. These hetero Diels-Alder cycloadditions can be very efficiently catalyzed by Lewis acids such as $ZnCl_2$ or $BF_3.Et_2O$ which, very probably, are mainly chelated to the oxygen of the P=O group, the electron-withdrawing effect of which is increased. Cycloadducts **24** can be selectively desulfanylated via a radical process (see Sect. 4) into phosphonodihydrothiopyrans **25**. Starting from diacetoxybutadiene, this sequence, cycloaddition-desulfanylation, has been applied to the synthesis of a diastereomeric mixture of diacetoxy

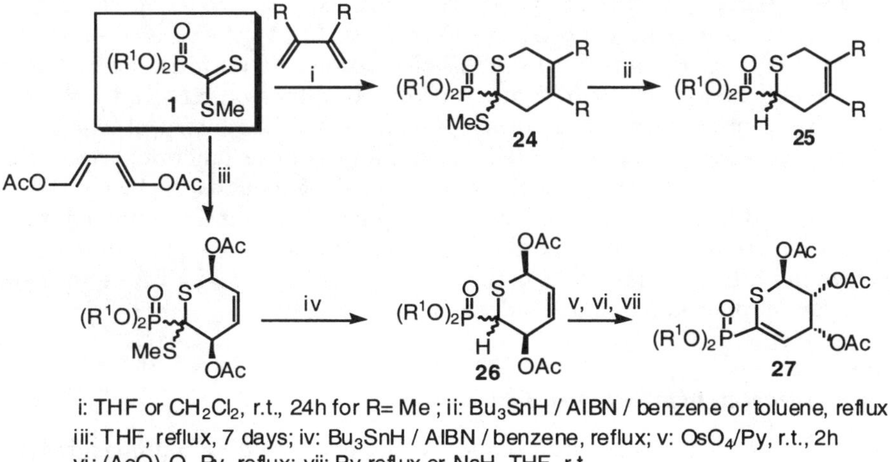

i: THF or CH_2Cl_2, r.t., 24h for R= Me ; ii: Bu_3SnH / AIBN / benzene or toluene, reflux
iii: THF, reflux, 7 days; iv: Bu_3SnH / AIBN / benzene, reflux; v: OsO_4/Py, r.t., 2h
vi: $(AcO)_2O$, Py, reflux: vii: Py reflux or NaH, THF, r.t.

Scheme 6

thiopyran **26**. After selective *cis*-dihydroxylation (*trans* to the acetoxy substituents), acetylation and elimination of one AcOH group in basic medium, the phosphonylated thiaglycoside **27**, a phosphono and thio analogue of a shikimic acid ester, has been obtained [18].

2.2 Phosphonodithioacetates

Two methods can be used to prepare phosphonodithioacetate **28** (Scheme 7). One method consists of the addition of methyl chlorodithioformates to the cuprate of a diethyl methylphosphonate [19], the other is the addition of HCl gas and EtSH to a diethyl (cyanomethyl)phosphonate followed by a sulfhydrolysis in pyridine [20].

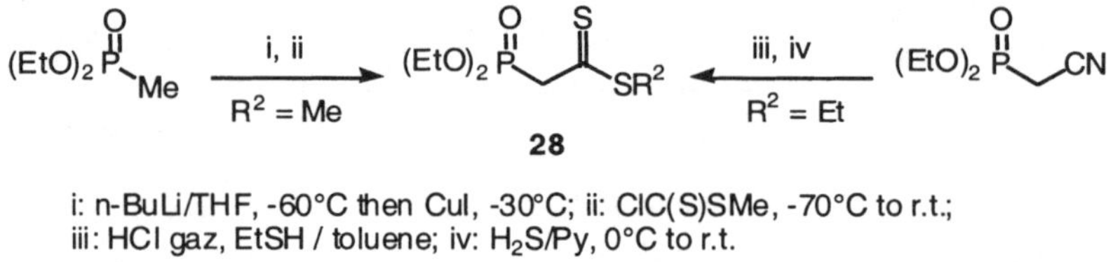

Scheme 7

In these type of dithioesters, the phosphonate function does not have a strong influence on the reactivity of the thiocarbonyl group. However, the two functions, which can be used successively in different reactions, are both associated with the high acidity of the methylene protons, making these compounds prone to reactions with carbonyl compounds in basic medium such as Horner-Wadsworth-Emmons (HWE) or Knoevenagel olefinations. Although, as reported below, phosphonodithioacetates such as **28** have been found very efficient in the Knoevenagel reaction, they are poor HWE reagents, as shown by the scarcity of described applications and modest reaction yields [19, 21]. Moreover, the α,β-unsaturated dithioesters obtained from this olefination reaction are not very stable if the double bond is not sufficiently substituted (dimerization or polymerization) [8a]. Three main recent developments of the chemistry of phosphonodithioacetates are herein reviewed – the synthesis of new functionalized HWE reagents via the thioacylation of amino substituted compounds, the hetero Diels-Alder cycloaddition of α,β-unsaturated phosphonodithioesters (obtained from **28** via Knoevenagel reactions), and the preparation and reactivity of fluorinated phosphonodithioacetates opening the way to the synthesis of a variety of new analogues of phosphorylated bioactive molecules.

2.2.1 *New HWE Reagents from Phosphonodithioacetates*

Thioacylation of primary and secondary amines, including aminoacids and aminoalcohols, has been readily carried out from phosphonodithioacetate **28** (R^2=Et) and the resulting phosphorylated thioamides (thiocarbamoylmethylphosphonates) have been applied to the olefination (with functionalization) of various aldehydic substrates (Scheme 8). HWE reaction can be performed using NaH/THF or K_2CO_3/H_2O [22]. Three interesting applications of this reaction are worth noting. The phosphorylated thioamides **29** derived from amino acids have been used for the synthesis of aminoacid-terminated dendrimers **30** (first, second, and third generations) [23]. From the phosphorylated thioamide **31**, prepared from piperidine, a *β*-D-mannofuranosyl-ethanethioamide **32** has been synthesized and converted into the corresponding dithioester **33**, the first C-glycoside derivative with *N*-thioacylating properties, in particular towards aminoacids [24]. From the hydroxy-substituted phosphorylated thioamide **34** obtained by thioacylation of homochiral aminoethanols, phosphonomethylthiazolines **35** have been prepared using the Mitsunobu procedure and these thiazolines have been used for the preparation of enantiopure *α*,*β*-unsaturated thiazolines **36** [25].

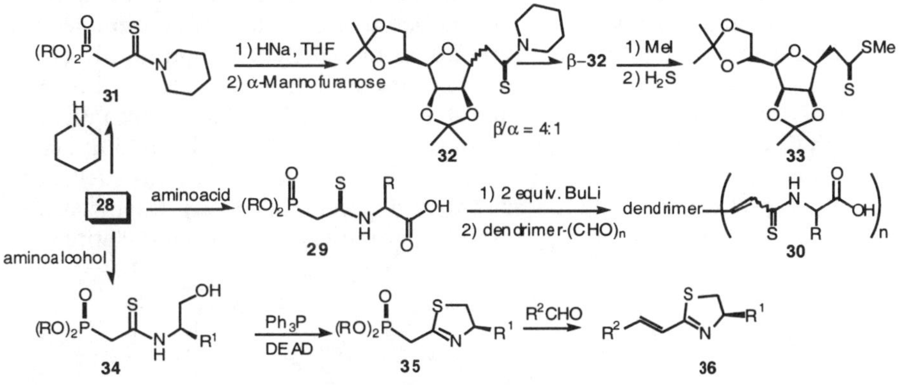

Scheme 8

2.2.2
Phosphonodithioacetate Derivatives as Heterodienes

Phosphonodithioacetate **28** has been found to be a good reagent for Knoevenagel reactions with aromatic aldehydes (or their aminals) and the corresponding heterodienes **37** (*E*) have been stereoselectively prepared in good yield [26a] (Scheme 9). At 125 °C or alternatively, at 20 °C under high pressure (11 kbar), inverse-electron demand hetero Diels-Alder reactions have been performed with enol and thioenol ethers leading to new phosphonosubstituted dihydrothiopyrans **38**. With ethyl vinyl ether (or thioether), an interesting inversion of the stereoselectivity of the reaction is observed between the thermal conditions (*exo*-approach of the dienophile) and the hyperbaric conditions, which favors the more compact transition state (*endo*-approach). Moreover, under high pressure, another switching of the stereocontrol of the cycloaddition is observed in the presence of pyridine. This is explained by a transient Michael-adduct triggering of an in-situ isomerisation ($E \rightarrow Z$) of the heterodiene **37** and a higher reactivity of the *Z* isomer [26b].

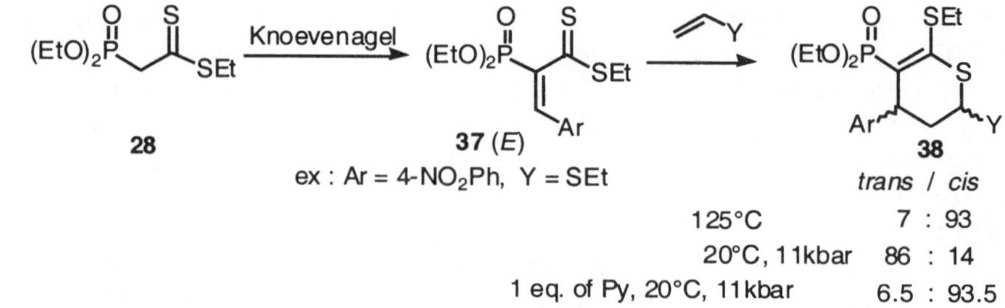

Scheme 9

2.2.3
Fluorinated Phosphonodithioacetates

The reactivity of difluoro phosphonodithioacetate **39** has been recently studied and its use as building block allows the preparation of a variety of new molecules functionalized by the difluorophosphono group which is a good mimic of the phosphate moiety of some bioactive molecules [27]. In dithioester **39**, the electron-withdrawing effect of the fluorine atoms also increases the electrophilic and dienophilic characters of the thiocarbonyl group. Thus, **39** has been found to be a good heterodienophile leading to dihydrothiopyranic compounds such as **40** and **41** [27a]. These cycloadducts are potential precursors of new thiaglycosides substituted by a difluoromethylphosphono group via their desulfanylation and *cis*-dihydroxylation (Scheme 10).

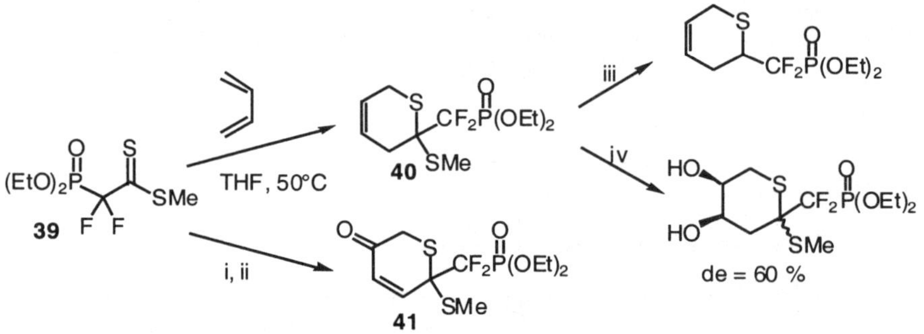

i: Danishefsky's diene; ii: TMSOTf ; iii: Bu_3SnH/AIBN; iv: $OsCl_3$ cat., $K_3Fe(CN)_6$, K_2CO_3 quinuclidine, H_2O/tBuOH.

Scheme 10

Fluorinated dithioester **39** has also been involved in thioacylation reaction of enantiopure aminoalcohols leading to the corresponding hydroxylated thioamides. The later, treated either by $SOCl_2$/pyridine, CH_3SO_2Cl, or Ph_3P/DEAD, cyclise into thiazolines **42** without epimerisation and in excellent yields. An interesting application of this reaction is the synthesis of a new type of modified nucleotides such as **43** and **44** by a one pot reaction simply by mixing the dihydroxy substituted thioamide (obtained from **39** and the 2-amino-1,3-

propanediol) with a nucleic base, using Mitsunobu's conditions. In this process, one first hydroxy group is involved in the formation of the thiazoline ring and the second is substituted by the nucleic base [27b] (Scheme 11).

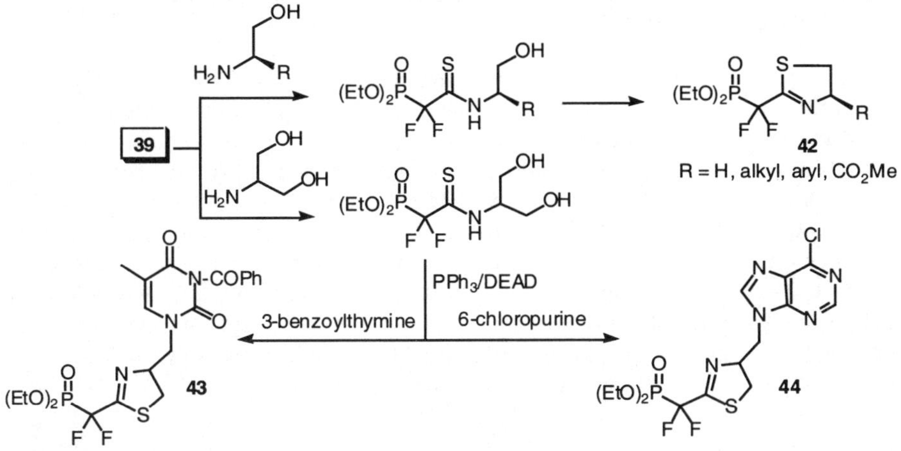

Scheme 11

During the thioacylation of less reactive amines by the phosphonodifluorodithioacetate **39**, a small amount (< 10%) of its monofluoro analogue has been detected. The formation of this by-product has been attributed to a thiophilic addition of the methanethiolate anion to **39** followed by a β-elimination of one fluorine atom [27b]. From this observation, a direct access to the monofluoro phosphonodithioacetate **45** by treatment of **39** with two equivalents of EtSLi has been recently achieved (Scheme 12). Compound **45** is a new P,S,F-substituted building block which, through applications similar to that performed for its difluoro analogue, is a precursor of a variety of new functionalized monofluoromethylphosphonates (such as monofluorinated analogues of **42**, **43**, and **44**). The latter are potential HWE olefination reagents, which can lead to new functionalized fluorovinylic derivatives [28].

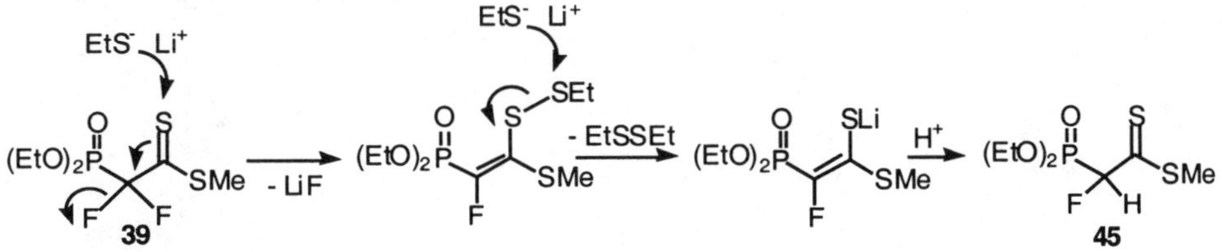

Scheme 12

3 Sigmatropic Rearrangements

3.1 [2,3]-Sigmatropic Rearrangement

The [2,3]-sigmatropic rearrangement of α-(allyloxy), -(allylamino) or -(allylthio) carbanions and ylides (also called Wittig, thia-, or aza-Wittig rearrangement) is a well-known process which has been widely used for regio- and stereocontrolled formation of carbon-carbon bonds. [29]. Numerous functional groups (aryl, alkenyl, alkynyl, or carboxylic moieties) have been used with the aim to stabilize the α-hetero-substituted carbanion, but it is only in the 1990s that this rearrangement has been studied in the phosphonate series. The reactivities of allyloxy- [30–33], allylamino- [34–37], and allylthio-methylphosphonate [38–42] carbanions and ylides have been systematically examined. It is interesting to note that the [2,3]-sigmatropic rearrangement has been observed only, with the carbanions for the O-allylic methylphosphonates, with the ylides for the N-allylic methylphosphonates, and with both carbanions and ylides in the sulfur series. Herein, are summarized the results obtained with allylthio-methylphosphonates, as an original preparation of new α-sulfanylphosphonates.

Thia-[2,3]-Wittig sigmatropic rearrangement of lithiated carbanions **47**, obtained by deprotonation of the S-allylic sulfides **46**, affords the thiols **48** or their alkylated derivatives **49**. The corresponding sulfonium ylides **51**, prepared by deprotonation of the sulfonium salts **50** also undergoes a [2,3]-sigmatropic shift leading to the same sulfides **49** [36, 38] (Scheme 13). As far as stereochemistry is concerned, with crotyl (R^1, R^3=H, R^2=Me) and cinnamyl (R^1, R^3=H, R^2=Ph) derivatives, it has been shown that the diastereoselectivity depends on the nature of the R^2 substituent and on the use of a carbanion or an ylide as intermediate.

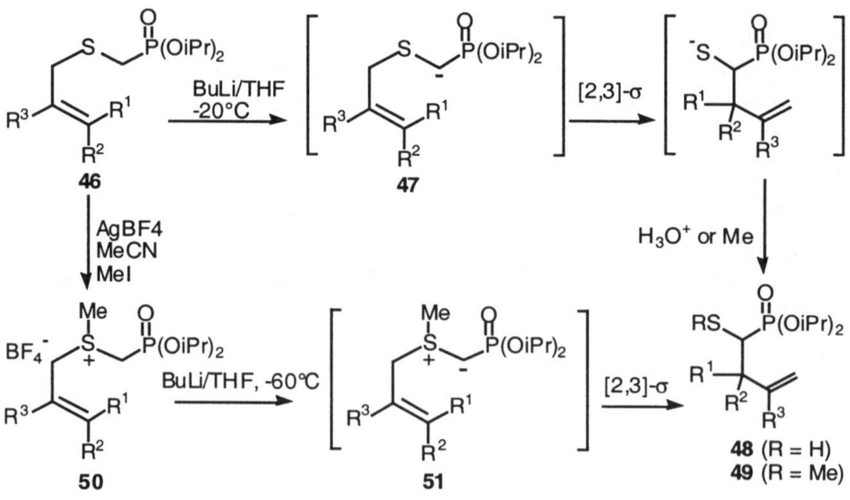

Scheme 13

The thiol **48** (with $R^1, R^2, R^3 = H$) undergoes intramolecular cyclisation in THF, in the presence of AIBN under UV irradiation, to give a 2-phosphonothiolane [36], a phosphorus and sulfur analogue of proline. More recently, an asymmetric version of the sequence, [2,3]-sigmatropic rearrangement and radical cyclisation, has been carried out (see Sect. 5.1.1.) [41].

The rearrangement has also been extended to the bis(phosphonate) substrate [37]. The failure of the carbanion derived from **52** to undergo a spontaneous [2,3]-σ shift is ascribed to its strong stabilization by the two adjacent electron-withdrawing phosphono groups. However, the sigmatropic shift leading to **56**, readily occurs from the S-allylic intermediate ylides **54**, resulting from the treatment of sulfonium salts **53** with sodium hydride. The rearranged products **56** can also be obtained starting either from the carbanion of the methylsulfanyl-methylene bis(phosphonate) **55** and allyl bromide or from the carbanion of **52** and methyl iodide. In both pathways, the assumed identical intermediate ylide **54** (resulting from a methylation or allylation of the sulfur atom instead of the α-anionic carbon) has not been detected (Scheme 14).

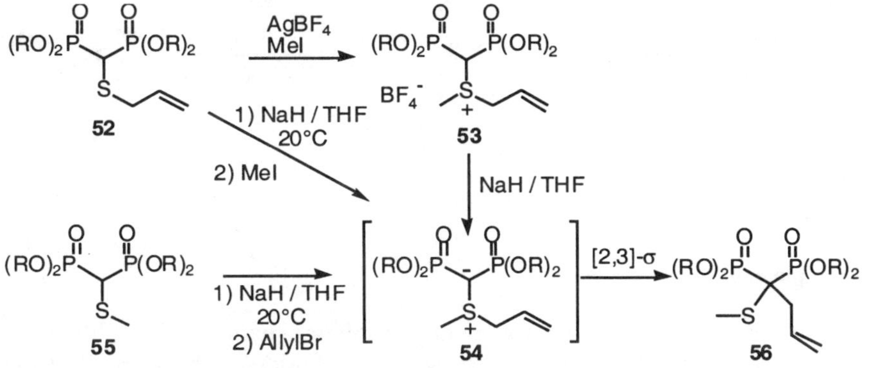

Scheme 14

A more direct access to the unstable and non isolated sulfonium ylides **58a–c** is the reaction of diisopropyl diazomethylphosphonate **57** with allylic sulfides, catalyzed by Cu(II), Rh(II) [39], or ruthenium porphyrins.[40] For example, the α-phosphorylated γ,δ-unsaturated sulfides **59–61** are obtained through the [2,3]-sigmatropic rearrangement of **58a–c**. This method allows the use of a greater variety of starting allylic sulfide substrates, such as 2-vinyl tetrahydrothiophene, or propargylic sulfides (Scheme 15).

A nice application of this reaction for the synthesis of cyclic α-sulfanylphosphonates **63** has been reported [42]. It involves a Rh(II)-catalyzed [2,3]-sigmatropic rearrangement and a ring-closing metathesis of the resulting α-(S-allyl) γ,δ-unsaturated phosphonates **62** (Scheme 16). However, the last step occurs with a low yield (19%) when $R^1 = H$.

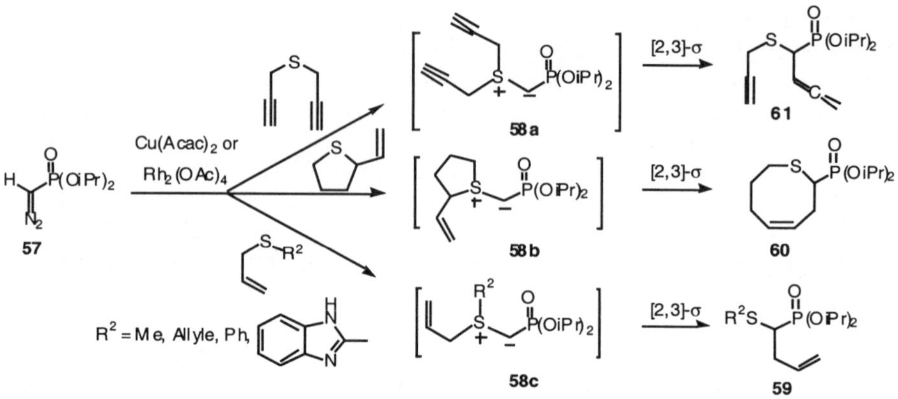

Scheme 15

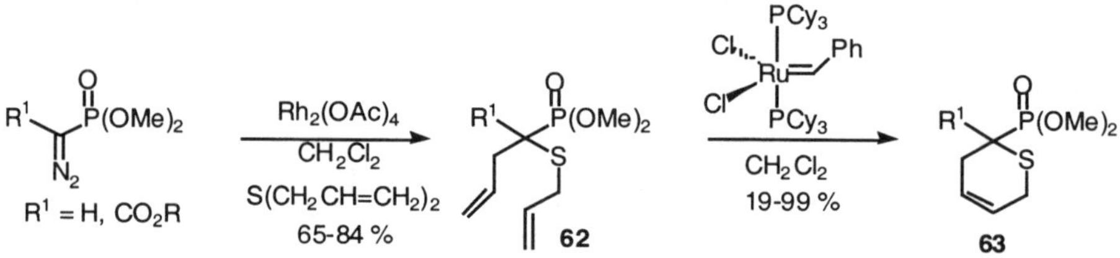

Scheme 16

3.2
[1,2]-Sigmatropic Rearrangement

In the oxygen series, the rearrangement phosphate↔hydroxyphosphonate involving a [1,2]-sigmatropic shift are known, on both sides, in basic media. However, in the case of the bis(phosphonate) derivatives, the conversion of hydroxymethylene bis(phosphonate) into the corresponding phosphonomethyl phosphate is usually favored (see references cited in [43]). In the sulfur series, the reverse rearrangement of lithium carbanion **65** (generated by deprotonation of the S-phosphonomethyl phosphorothiolate **64**) into the thiolate bis(phosphonate) **66** readily occurs at –35 °C [43]. Although thiolate **66** can readily loose sulfur at higher temperature (to give the methylene bis(phosphonate) carbanion), the corresponding thiol and various sulfides **67** have been prepared by careful protonation or addition of alkylating agents at – 40 °C (Scheme 17).

The [1,2]-σ rearrangement of phosphinothioates into (alkylsulfanylmethyl)phosphine oxides using a chiral phosphinoyl group has also been reported (see Sect. 5.1.1.).

A recent paper [44] shows that the treatment of silyl thioketones **68** with lithium diethylphosphite proceeds via a thiophilic attack followed by a thiophosphate → mercaptophosphonate (**69**→**70**) carbanionic rearrangement and the migration of the silyl group from the carbon to the sulfur atom leading to the S-silylated sulfanylphosphonate carbanion **71**. The last step represents the first example of the thia-Brook rearrangement (Scheme 18).

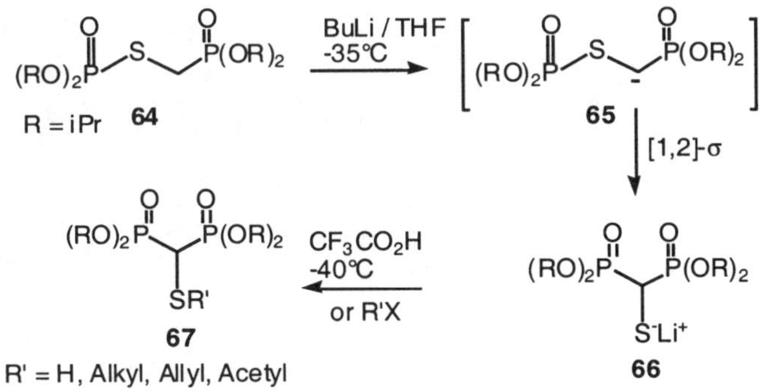

Scheme 17

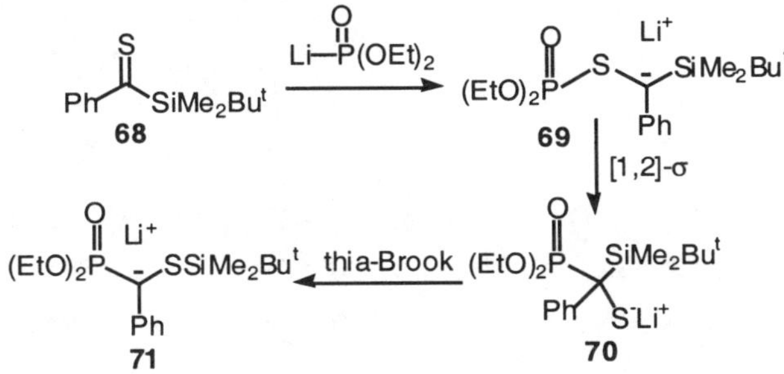

Scheme 18

3.3
[1,3]-Sigmatropic Rearrangement

The oxygen to carbon 1,3-migration of the phosphoryl group (also called arylphosphate – *o*-hydroxyarylphosphonate rearrangement) associated with the *ortho*-lithiation of diethyl aryl phosphates by LDA has been published and developed in the 1980s [45]. First studies related to the transposition of this rearrangement in the thiophosphate series have been unsuccessful due to a predominant attack of the base to the P=O group and cleavage of the P–S bond leading to thiophenol [46]. The substitution of the ethyl substituents of the phosphoryl moiety by more bulky isopropyl groups overcomes this restriction and *o*-sulfanyl arylphosphonates **72** or other aryl- or heteroaryl-phosphonates presented in Scheme 19 have been obtained in good yield via the *o*-lithiation-[1,3]-sigmatropic rearrangement. This process has then been extended to S-heterocyclic phosphorothiolate and thio-substituted pyridyl and thienyl phosphonates have been prepared [47]. Recently, from the (*o*-methylsulfanylphenyl) phosphonate, new heterocycles **74** and **75** have been synthesized [48a]. The five-membered ring **74** is obtained by a nucleophilic attack of the sulfur stabilized carbanion of **73** on the phosphoryl group activated by a Lewis acid. The six-membered ring **75** results from the monohydrolysis of the phosphoryl group of sulfide **73**, oxidation of the latter into sulfoxide, and then a Pummerer type reaction in-

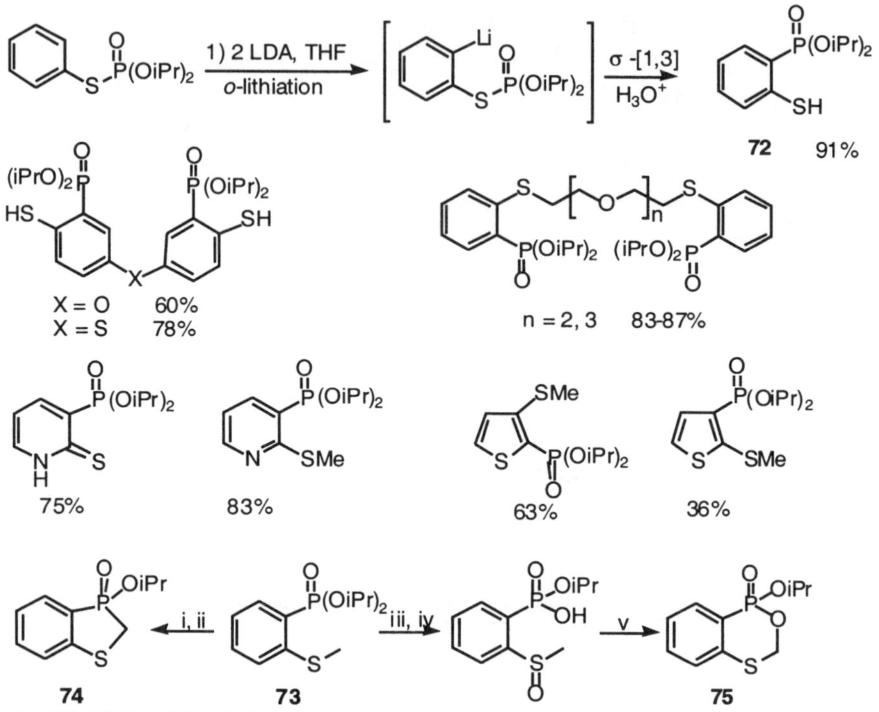

i: LDA, THF; ii: $BF_3.Et_2O$ or $ZnCl_2$, 1 equiv.
iii: NaN_3, DMF, 100°C; iv $NaIO_4$, acetone, H_2O; v: $(CH_3CO)_2O$

Scheme 19

volving the participation of the oxygen of the phosphonic function as nucleophile. A first example of the chelating properties of the *ortho*-P,S-difunctional compounds has been the isolation and characterization of a new dimeric platinum complex from sulfide **73** [48b]. Besides, from **73** (R=Me), an enantiopure (*S*)-(–)-sulfoxide (*o*-methylsulfinyl phenylphosphonate) has been prepared (see Sect. 5.2.1).

The same type of *o*-lithiation-sigmatropic rearrangement has also been observed from *N,N,N′,N′*-tetramethylphosphorodiamidothiolate **76** with *sec*-BuLi as the base and addition of methyl iodide. The latter reagent alkylates the assumed arylthiolate formed by the S–C migration of the phosphoramido group (Scheme 20). However, the thiol corresponding to the rearrangement of **76** has not been prepared [49]. From the phosphorodiamidothiolate **77** derived from (*S*)-(+)-2-(anilinomethyl)-pyrrolidine, the same *o*-lithiation by LDA followed by protonation, leads to the expected *o*-sulfanyl derivative but in a low yield (20%) [50]. Recently, it has been shown that in the same conditions, no 1,3-migration occurs from the *o*-lithiated chiral phosphorodiamidothiolate **78**, unless an alkylating agent or a Lewis acid is added to the *o*-lithiated intermediate [48a]. Therefore, the use of such reagents promoting the rearrangement should allow its generalization to a variety of chiral phosphorodi-

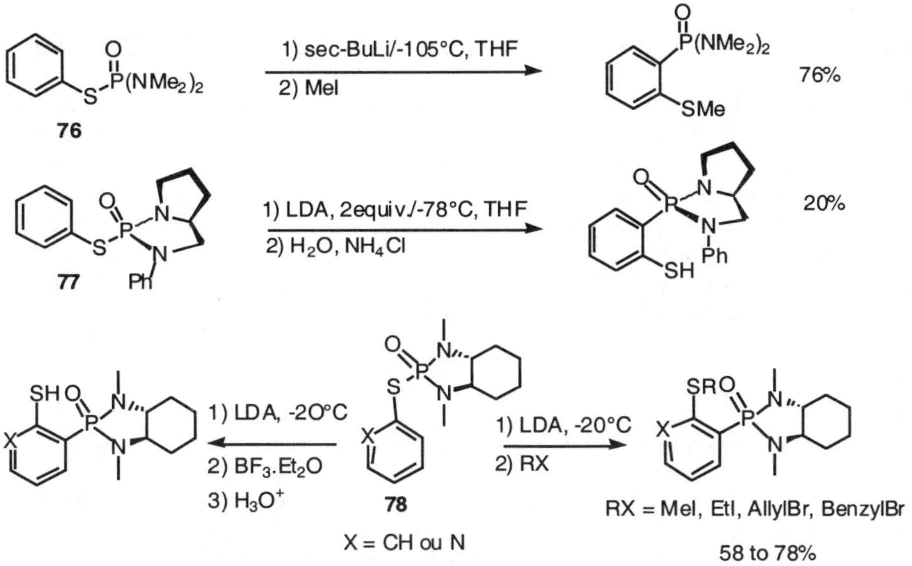

Scheme 20

amidothiolates, which could be prepared from various enantiopure C_2-symmetric diamines.

The rearrangement has also been extended to P-chiral S-phenyl phosphinothiolate **79** and *O*-phenyl phosphinothioate **80** (Scheme 21). With these asymmetric compounds, the C–P bond formation was found to occur stereoselectively and with retention of configuration at phosphorus [51].

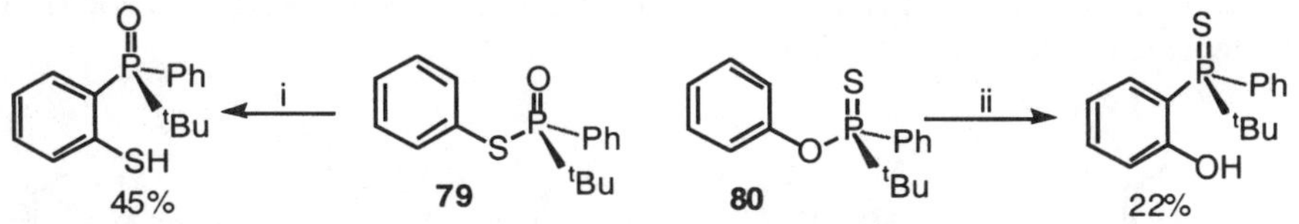

i: LDA, 4equiv., THF, -78°C, 30min then H_3O^+; ii: t-BuLi, 5 equiv.-78 to 20°C, THF then H_3O^+

Scheme 21

All these results show that the *o*-lithiation followed by [1,3]-sigmatropic rearrangement is a valuable process creating an aromatic C–P bond and providing access to new difunctional chelating compounds. Owing to the possibility of introducing asymmetry both at the sulfur or phosphorus sites, these bidentate ligands should find applications in catalytic asymmetric reactions as it has been already exemplified with some of their oxygenated analogues [52].

4
α-Thio-Substituted Phosphonates as Synthetic Tools in Radical Chemistry

Both organosulfur and organophosphorus compounds are known as important reagents in radical chemistry: radical acceptors (vinylphosphonates, vinylsulfoxides or sulfones, thiocarbonyl derivatives), heteroatom stabilized free radicals (centered on sulfur, on phosphorus, or on carbon atom bearing a sulfanyl or phosphoryl group), or precursors of radicals by cleavage of a carbon-heteroatom bond. This is the main reason why α-thiosubstituted phosphonates have emerged over the past few years as interesting tools in radical-based chemistry. Some typical reactions and their applications in organic synthesis are here reported.

4.1
Cleavage of the C–S bond α to a Phosphoryl Group

Barton et al. [53, 54] have developed a new radical based methodology for the synthesis of sugar and nucleoside phosphonates, analogues of the corresponding monophosphates. Using the ester of *N*-hydroxy-2-pyridinethione such as the thiohydroxamic uronic ester **81**, the method consists of the generation of a tetrahydrofuranic carbon centered radical. Addition of this radical to a vinyl phosphonate leads to a new phosphorylated C-radical, which is trapped by pyridinethione. Removal of the pyridinesulfanyl residue by homolytic cleavage of the C–S bond using Bu_3SnH/AIBN and further functional group transformations lead to the nucleotide analog **82** which mimics AZT-5′ monophosphate (Scheme 22) [53]. This chemistry has also been used to prepare a protected 6′-(pyridin-2-ylthio)-homouridine phosphonate which is then oxidized to the sulfone, fluorinated in α position of phosphorus with Selectfluor to give, after desulfonylation and deprotection, the first reported 6′-deoxy-6′-fluorohomonucleoside 6′-phosphonate [55].

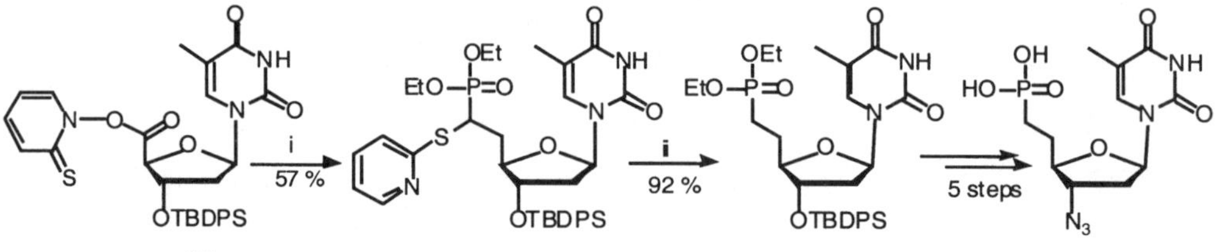

81 **82**

i: $(EtO)_2P(O)CH{=}CH_2$, hv, 0°; ii: Bu_3SnH, AIBN, C_6H_6 under reflux.

Scheme 22

In another version of this method, the radical generated by radical exchange from the aryl telluride carbohydrate **83** and the *N*-acetoxy-2-thiopyridone affords, after intramolecular cyclization and desulfanylation, the polyhydroxylated and phosphorylated pseudo sugar **84** [54] (Scheme 23).

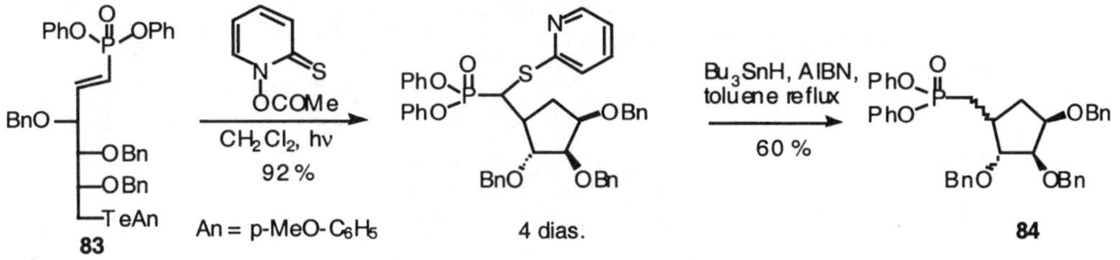

Scheme 23

A general study of the α-desulfanylation (and deselanylation) of phosphonates using Bu_3SnH/AIBN has been carried out [56]. The reduction turns to operate through the attack of the tin radical on the sulfur atom and generally requires two equivalents of Bu_3SnH. The cleavage occurs in the case of α-phosphoryl sulfides, but not for the sulfoxides and sulfones derivatives. Chemoselectivity has been studied in the case of various phosphonates additionally functionalized in the α-position. Main results show that sulfanyl groups can be selectively removed from α-phosphoryl O,S- and S,S-acetals (phenylthio is reduced in preference to methylthio), and also from alkyl-, aryl-, or ketophosphonates. Moreover, the reduction of halogen tolerates the presence of the sulfanyl group (Scheme 24).

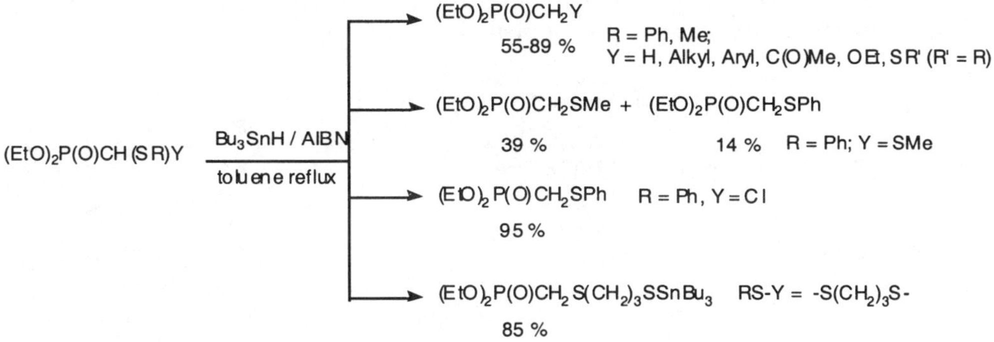

Scheme 24

The method has then been efficiently used to remove selectively the exocyclic methylsulfanyl group from phosphorylated thiopyranyl derivatives **85** resulting from hetero Diels-Alder reaction of a phosphono-dithioformate or dithioacetate (see Sects. 2.1.2 and 2.2.3). Functionalized thiopyrans **86** [17, 18, 27a] are thus obtained (Scheme 25). Owing to this selective desulfanylation, phosphonodithioesters can be used as heterodienophiles in place of the corresponding phosphonothioaldehyde, not described so far and probably very unstable.

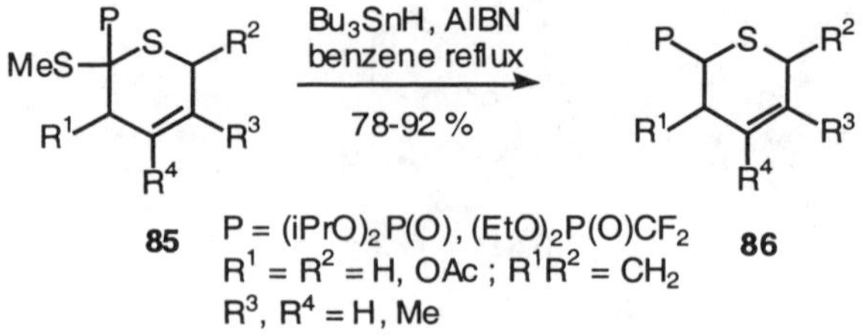

Scheme 25

4.2 Generation and Addition Reactions of Phosphorylated C-Radicals

The synthesis of α-substituted phosphonates **89**, via the electrophilic addition of phosphorylated C-radicals **88** (generated by reaction of Bu_3SnH to the readily accessible α-phosphoryl sulfides (or selenides)) and electrophilic addition to electron rich alkenes, has been described [57] (Scheme 26). A large excess of alkene is necessary to minimize the competitive formation of the undesired compound **90** resulting from direct reduction of the initial radical **88**. The ratio **89/90** has been measured for each example. The synthesis of the α-mono- or α,α-disubstituted (R^1 or $R^2 \neq H$) phosphonates **89** shows that the free radical approach complements the commonly used Arbuzov reaction. This methodology has been nicely illustrated by the synthesis of methylenomycin B [58].

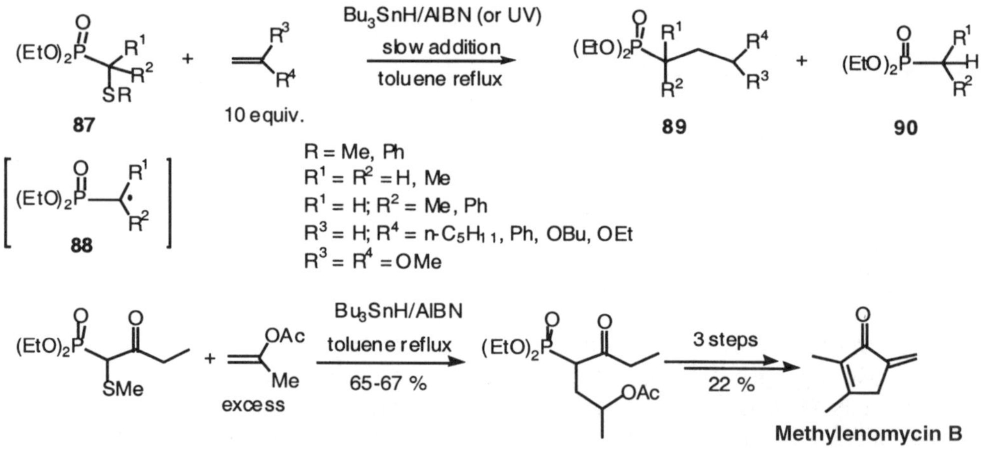

Scheme 26

Another interesting application is the C–S (or C–Se) homolytic cleavage from (α,α-difluoro-α-sulfanyl)methylphosphonates **91** (and selanyl analogues) which generates a phosphonodifluoromethyl radical (Scheme 27). A large variety of unactivated and electron-rich alkenes has been successfully used as radical acceptors to give new difluoromethylphosphonates derivatives **92–95**, analogues of the

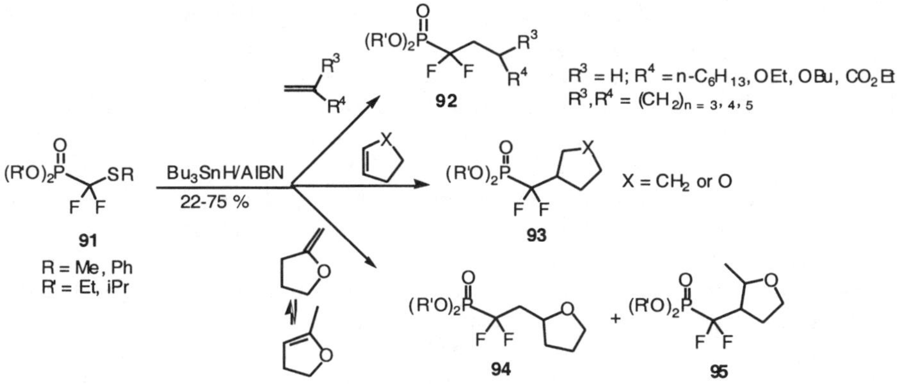

Scheme 27

corresponding phosphates [59]. One electron-poor alkene (ethyl acrylate) leads also to the expected addition product in 33% yield, proving both the electrophilic and nucleophilic character of the phosphonodifluoromethyl radical.

4.3
Phosphonodithioformates as Radical Trapping Agents

The preferential thiophilic attack of nucleophiles (organometallics, trialkylphosphites, thiolates) to the thiocarbonyl group of phosphonodithioformates has been highlighted in Sect. 2.1.1. The behavior of these dithioesters towards radical addition has also been investigated [60–64]. The reaction of methyl (diethylthiophosphono)dithioformate **96a** and methyl (diethylphosphono)dithioformate **96b** with a number of alkyl, alkoxy, thiyl, and organometallic radicals species (Scheme 28) results in the regiospecific thiophilic addition leading to rather persistent spin adducts **97**. The only exception is provided by manganese decacarbonyl, which only reacts with **96a** to give a binuclear adduct **98**. All the radicals have been characterized by EPR (electron paramagnetic resonance) spectroscopy [60].

The investigation has then been extended to triphenymethyl-, allyl-, benzyl- and but-3-enyl (diethylphosphono)dithioformates [61]. Besides, the transient

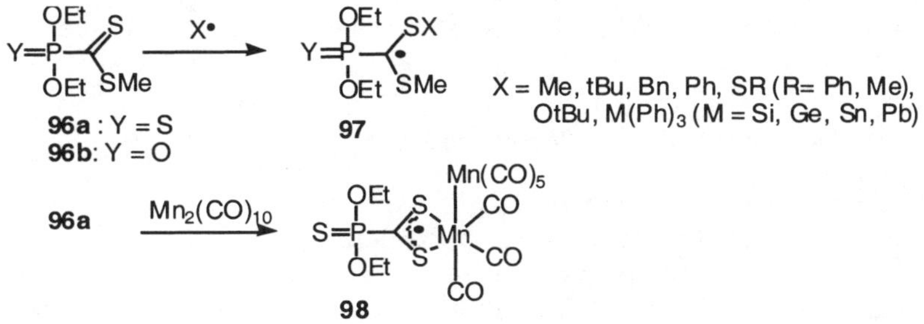

Scheme 28

EPR signals observed for the reaction of these dithioesters with a Grignard reagent is attributed to a radical anion formed by a single electron transfer from the organometallic to the thiocarbonyl group. Moreover, the same radical anions have been generated by electrochemical reduction of the phosphonodithioformates and studied by voltammetry and EPR spectroscopy [62]. Diarylphosphono- and di(2-menthyl)phosphono-dithioformates exhibit an excellent spin trapping ability (proved in ESR experiments), the latter yielding a radical adduct particularly persistent because of the steric hindrance provided by the bulky menthyl groups. Their remarkable property to act as process stabilizers, tested in the extrusion of polypropylene is also demonstrated [63a]. Benzyl (diethylphosphono)dithioformate and its thiophosphono analogue have been studied as Reversible Addition Fragmentation Transfer (RAFT) agents for a controlled radical polymerization of styrene [63b]. More recently, the reactivity towards free radicals of phosphono thiolformates and thionformates has also been examined [64]. Spin adducts are also observed by EPR for the thionesters which, are remarkably less efficient radical traps than the corresponding dithioesters.

5 Chiral Phosphorus and Sulfur-Containing Compounds in Asymmetric Synthesis

The wide application of chiral P,S-containing molecules as tools in asymmetric synthesis are closely linked to the strategies which allow the creation of a stereogenic center on phosphorus, on sulfur, or on the carbon bearing these heteroatoms. The following section summarizes recent methods described for the generation of such asymmetric difunctional compounds (α- and β-thio-substituted organo-phosphorus compounds, chiral P,S-ligands) and some of their applications. More often, in the selected structures the sulfur atom is on a carbon α or β to the phosphorus.

5.1 α-Thio-Substituted Organophosphorus Compounds

Chiral phosphoryl and sulfinyl groups are known as efficient auxiliaries in asymmetric synthesis. As reported below, their asymmetric induction in the α-position has been used to prepare chiral non-racemic organophosphorus compounds α-substituted by a sulfur function. Such compounds can also be obtained from their α-hydroxy analogues by OH$\rightarrow$SR stereoselective transformation.

5.1.1 *Asymmetric Induction of the Chiral Phosphoryl Group*

Starting from the (allylsulfanyl)methylphosphonate **99** and using dimenthylphosphonyl ester as chiral auxiliary in carbanionic [2,3]-sigmatropic rearrangement (see Sect. 3.1), α-mercaptophosphonate **100a** and sulfide **100b** have been obtained with high diastereoselectivity [65]. Radical intramolecular *5-endo*-cyclization of thiol **100a** leads to the corresponding thiolane **101** without epimerization (Scheme 29). The oxidation of 2-phosphonothiolane 3 into the corresponding sulfoxide **102** is totally stereoselective with a complete transfer of the asymmetry from the chiral α-carbon to the sulfur. The relative *trans* configuration of the C–P and S–O bonds has been demonstrated. It is interesting to note that the same thiolane **101** has also been prepared from thiolane S-oxide and trimenthylphosphite using a new Pummerer-phosphorylation reaction, but no diastereoselectivity was observed in this process [66].

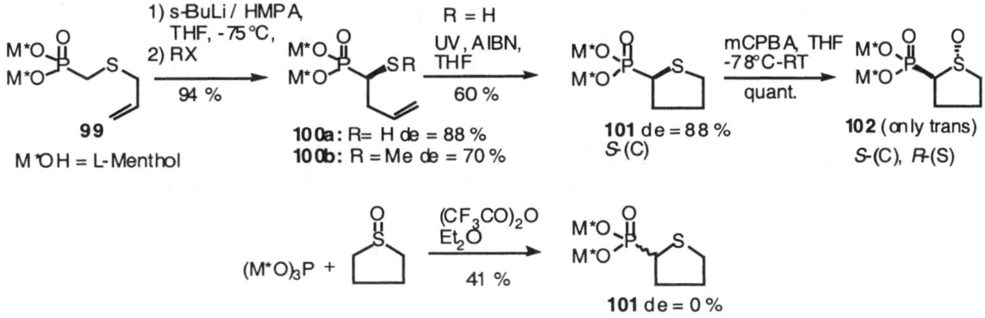

Scheme 29

In the asymmetric version of the [1,2]-σ Wittig rearrangement (see Sect. 3.2), the deprotonation of S-methyl (*tert*-butyl)arylphosphinothioate **103** followed by alkylation affords the corresponding (alkylthiomethyl)phosphine oxides **104** together with over-reacted products **105** (no diastereomeric excess is observed for this compound) and **106** [67] (Scheme 30).

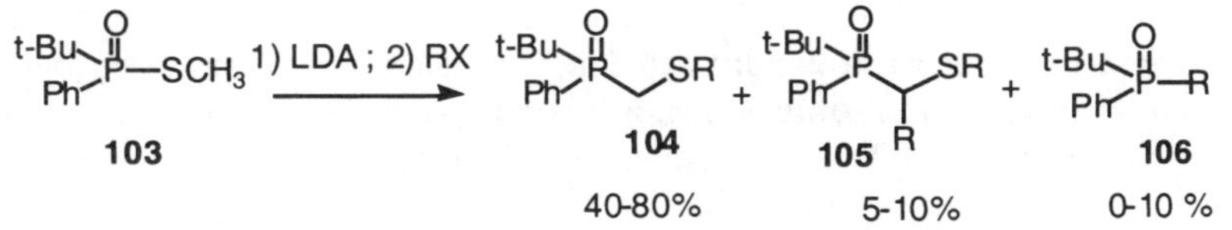

Scheme 30

In their work concerning the anions derived from chiral phosphonamidates, Denmark and Chen [68] obtain a high diasteromeric excess of 88% for the compound **108** resulting from the addition of the carbanion of phenylsulfenyl phosphonamidate **107** to 4-*tert*-butylcyclohexanone (Scheme 31).

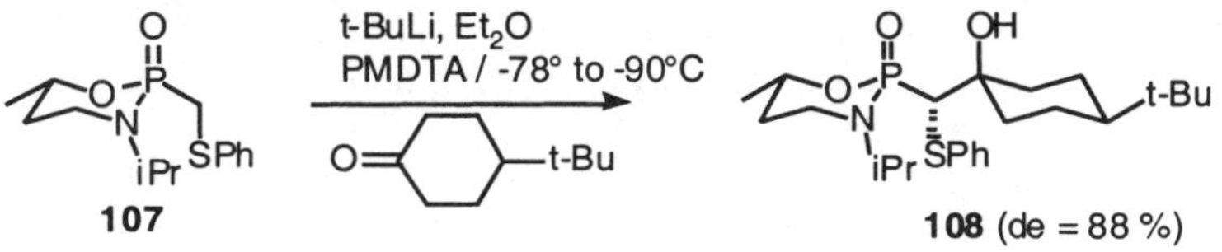

Scheme 31

α-Phosphono sulfonate **109** (BMS-188494) is a potent squalene synthase inhibitor [4b]. Both enantiomers have been prepared using a diastereoselective alkylation or sulfanylation of the C2-symmetric phosphonamide substituted carbanions of **110** (pathway A) and **111** (pathway B) [69] (Scheme 32).

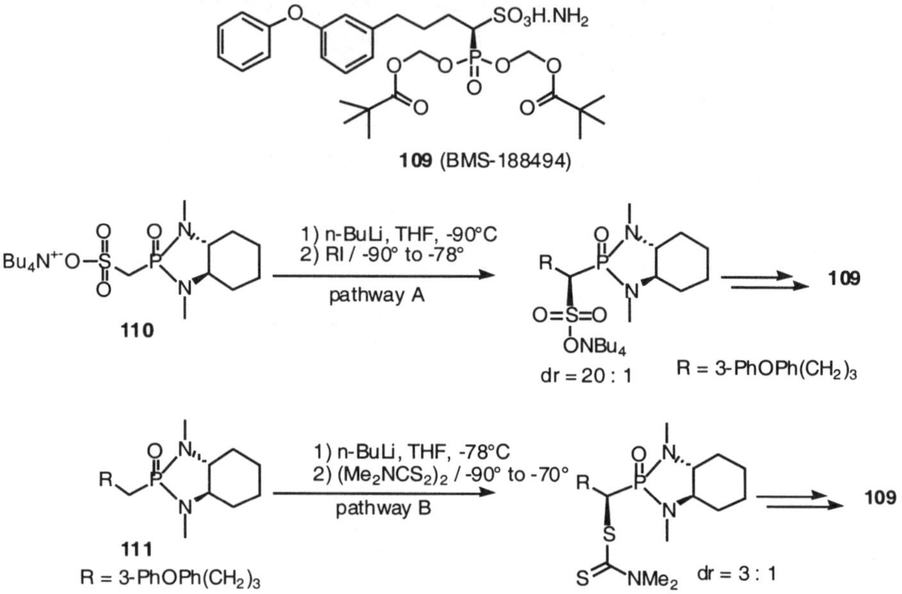

Scheme 32

A stereoselective radical reaction of the ester of *N*-hydroxy-2-pyridinethione (see also Sect. 4.1) to chiral vinylphosphine oxides has also been described and moderate to good diastereomeric ratios have been obtained for the compound **112** [70] (Scheme 33).

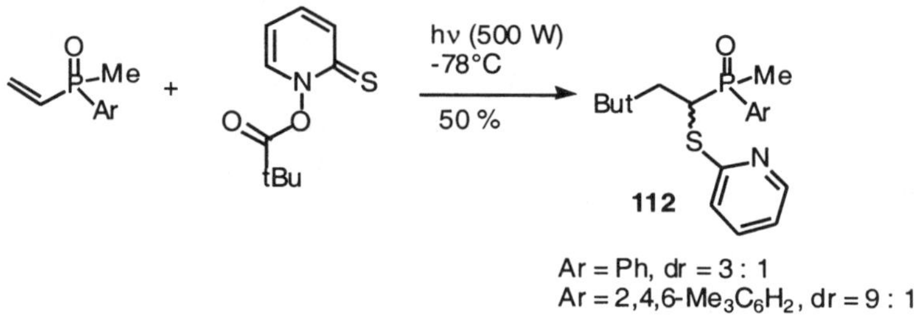

Scheme 33

5.1.2
Asymmetric Induction of the Sulfinyl Group

Among bifunctional compounds bearing phosphorus and sulfur centers, the α-phosphoryl chiral sulfoxides are undoubtedly the most widely used, in particular for the insertion of the sulfinyl group (used then as a chiral auxiliary) in a substrate by Horner-Wadsworth-Emmons (HWE) olefination. The very large number of examples does not allow us to include this reaction in the present review. Many studies of the asymmetric induction using chiral nonracemic α-phosphoryl sulfoxides concern especially the generation of a stereogenic carbon center α to the sulfinyl group. One of the most commonly used method for the preparation of chiral α-phosphoryl sulfoxides **113**, described by Mikolajczyk et al., consists of the addition of methylphosphonate lithium carbanions to the enantiopure menthyl *p*-toluenesulfinate [71]. The reactivity of enantiopure **113** has been widely examined and then involved in various synthetic applications. For example, the stereochemical study of the α-chlorination of this compound shows that the reaction occurs with prevailing retention of the configuration at the sulfur atom and the diastereomeric ratio at the chiral carbon is approximately 3:1, with the less thermodynamically stable diastereomer slowly isomerizing into the more stable one **114** [72]. The enantiomeric excess of the acetoxy-substituted phosphonate **115**, obtained from **113** by the Pummerer reaction, can be improved by replacing acetic anhydride [73] with ethoxy vinyl acetate [74] (Scheme 34). The sulfanylation of **113** with S-methyl methanethiosulfonate leads to the phosphonoformate dithioacetal mono-S-oxides **116** with a moderate diastereomeric ratio [75].

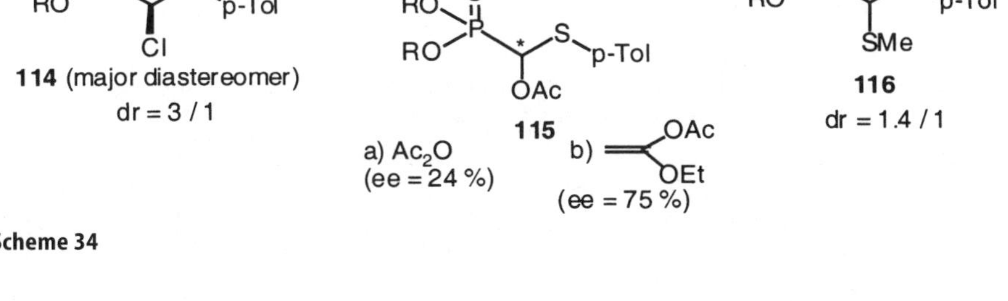

Scheme 34

An α-fluorinated phosphinyl sulfoxide derivative, prepared as a mixture of diastereomers by the treatment of lithiated (α-fluoromethyl)diphenylphosphine oxide with menthyl *p*-toluenesulfinate, has been used for the preparation of enantiomerically pure 1-fluorovinyl and 1-fluoromethyl sulfoxides [76]. (*R*)-Diethyl *tert*-butylsulfinylmethylphosphonate **117** has been prepared by the sulfinylation of the lithium carbanion of diethyl methylphosphonate with the (*R*)-*tert*-butyl *tert*-butanethiolsulfinate. Sulfoxide **117** has then been used to introduce the chiral sulfinyl group as a chiral auxiliary in sulfinylated enynes, which are substrates for asymmetric Pauson-Khand reactions [77] (Scheme 35).

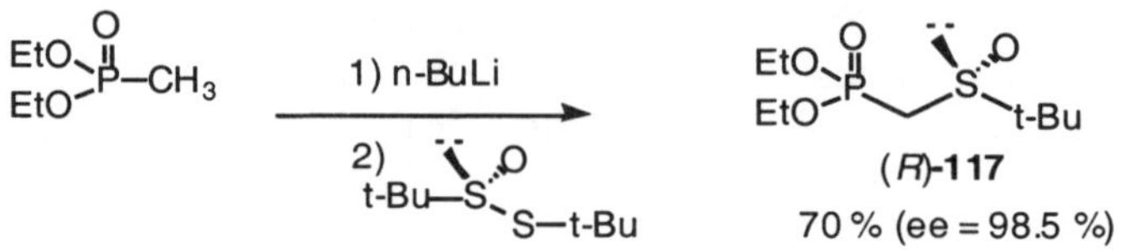

Scheme 35

An alternative synthesis of the chiral non-racemic α-phosphoryl sulfoxides is the enantio- or diastereoselective oxidation of the corresponding sulfides. For example, asymmetric oxidation using (+)-8,8-dichlorocamphorylsulfonyloxaziridine, cumene hydroperoxide in the presence of titanium/diethyl tartrate (DET) complexes or involving the asymmetric induction of a chiral phosphorus bearing one menthyloxy group, have been used for this purpose [78]. Resulting α-phosphoryl sulfoxides **118** and **119** are precursors of enantiopure dialkyl sulfoxides **120** via their reaction with Grignard reagents, the phosphonomethyl moiety being used as the leaving group (Scheme 36).

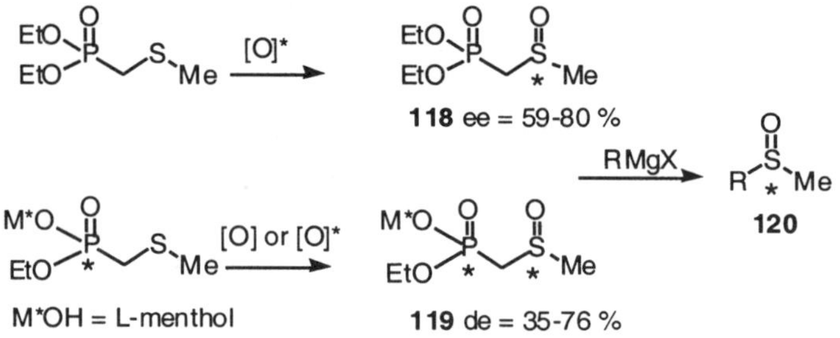

Scheme 36

Oxidation of cyclic phosphonoformaldehyde dithioacetal, using the Modena protocol, yields the trans disulfoxide **121** in excellent enantiomeric excess. Then **121**, via HWE olefination and oxidation of the double bond has been used for the diastereoselective preparation of spirocyclic *bis*-sulfinyl oxiranes (new versatile intermediates in asymmetric synthesis) [79] (Scheme 37).

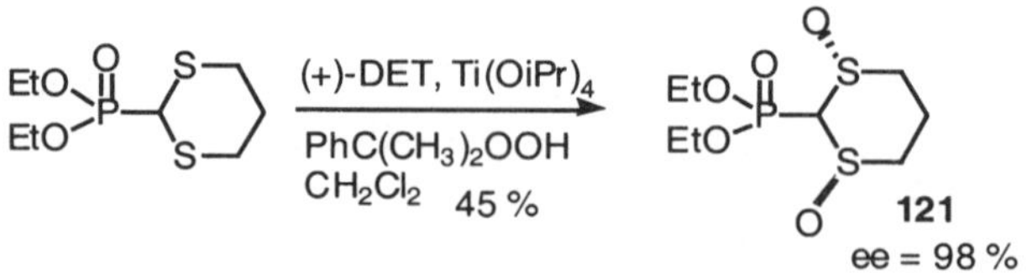

Scheme 37

The optically active α-sulfinyl vinylphosphonate **122** prepared in two different ways (Scheme 38) is an interesting reagent for asymmetric synthesis [80]. This substrate is an asymmetric dienophile and Michael acceptor [80a]. In the Diels-Alder reaction with cyclopentadiene leading to **123**, the endo/exo selectivity and the asymmetry induced by the sulfinyl group have been examined in various experimental conditions. The influence of Lewis acid catalysts (which also increase the dienophilic reactivity) appears to be important. The 1,4-addition of ethanethiol gives **124** with a moderate diastereoselectivity.

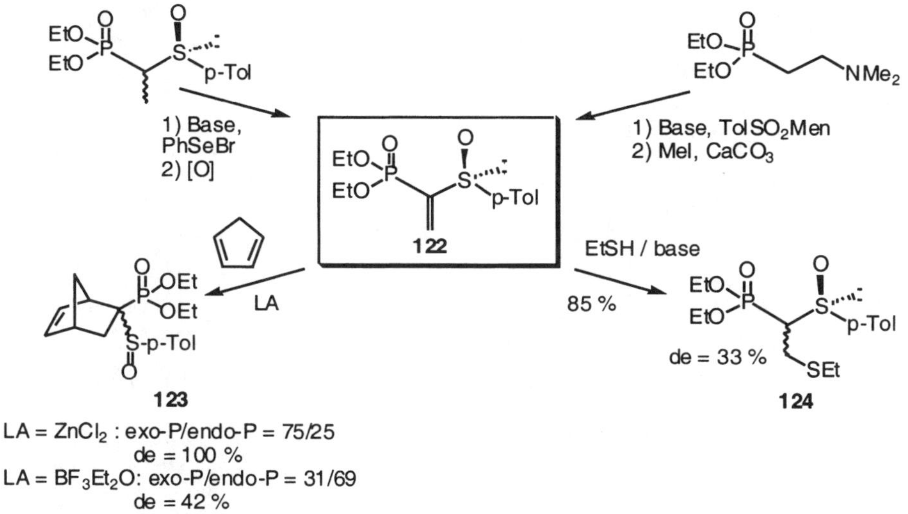

Scheme 38

The 1,3-dipolar cycloaddition of diazoalkanes to enantiopure sulfoxide **122** (or homologues monosubstituted on the double bond) has been described as a new synthetic route to 3-phosphorylpyrazoles such as **125**. In contrast, using diphenyldiazomethane, the reaction occurs with elimination of nitrogen leading to the chiral phosphoryl and sulfinyl cyclopropane **126** with total diastereoselectivity [81] (Scheme 39). The cyclopropane derivatives **127** (R=D) and **128** (R=Me) have been prepared using respectively the fully deuterated dimethylsulfoxonium methylide and the diphenylsulfonium isopropylide [82a]. More recently, *E*-(*S*)-(1-diethylphosphono-2-phenyl)vinyl *p*-tolyl sulfoxide **122b** has been found to undergo cyclopropanation with three sulfur ylides (dimethoxy(oxo)sulfonium methylide, diphenyl sulfonium isopropylide, and ethyl(dimethylsulfuranylidene)acetate) in a highly diastereoselective manner. From the major cyclopropanic diastereomer obtained with the third ylide, an enantiopure constrained analogue of the $GABA_B$ antagonist, phaclofen, has been synthesized [82b].

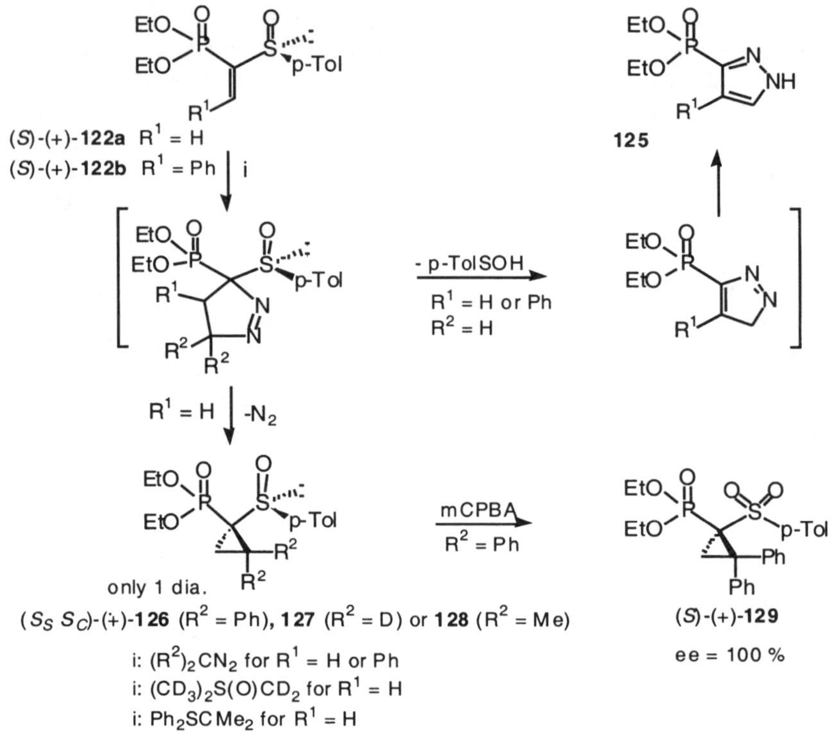

Scheme 39

5.1.3
Stereoselective OH → SR Transformation

Compared to α-amino- or α-hydroxyphosphonates for which various asymmetric syntheses have been proposed, no general and efficient method giving access to optically active α-sulfanylphosphonates (or derivatives) have been described until recently. One example of the asymmetric synthesis of an α-sulfanylphosphonate is given by the preparation of (1-sulfanyl-but-3-enyl) phosphonate **100** from a [2,3]-sigmatropic rearrangement (Sect. 5.1.1). Another example is the Mitsunobu reaction from *R*-α-hydroxyphosphonate **130** which gives **131** with a high diastereoselectivity but in low yield [65] (Scheme 40).

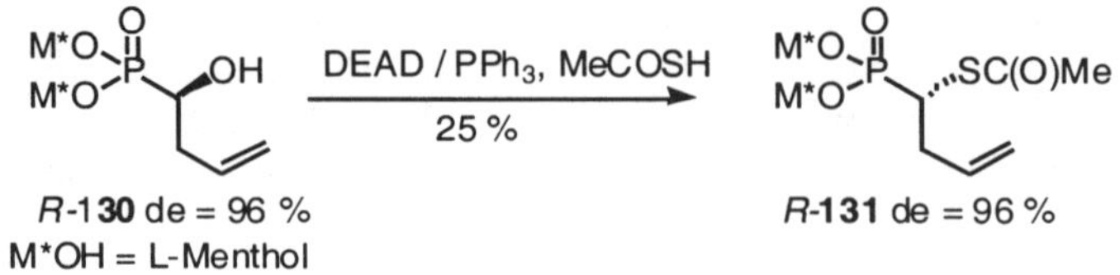

Scheme 40

An S-α-acetoxyphosphonate, potential precursor of the biologically active BMS-188494 **109** (see Sect. 5.1.1, scheme 32) has been prepared by lipase-catalyzed stereoselective acetylation [83]. However, to our knowledge, its conversion into the corresponding S-α-phosphono sulfonic acid has not been performed so far. Very recently, an efficient and general method allowing the conversion of α-hydroxyphosphonate to the corresponding sulfanyl analogues has been proposed [84] (Scheme 41). Thus, starting from enantiopure α-hydroxyphosphonates **132** obtained by enzymatic resolution, optically active α-sulfanylphosphonates **135a** and their corresponding methylsulfides **135b** have been prepared in three stereoselective transformation steps involving a *p*-nitro-tosylate **133** and a thiocyanate **134** as intermediates. An important point to note in this process, in order to avoid epimerization during the reducing cleavage of **134**, is the addition of two equivalents of trimethylchlorosilane or methyl iodide together with the lithium borohydride.

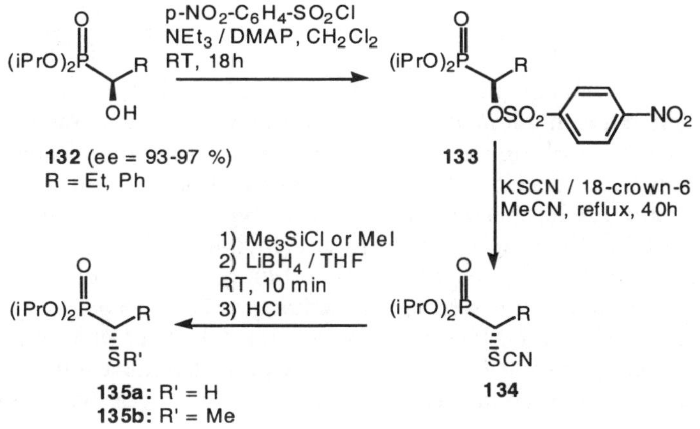

Scheme 41

5.2
β-Thio-Substituted Organophosphorus Compounds

Among β-thiosubstituted organophosphorus compounds bearing chiral groups, phosphono methyl thiazolines (Sect. 2.2.1, Scheme 8) and *o*-sulfanyl aryl phosphonamides or phosphinoxides (Sect. 3.3, schemes 20 and 21) have already been mentioned. As a complement to this, some recent synthesis of non racemic β-sulfinyl phosphines and phosphonates and thiazolidinyl phosphonates are reported below. Moreover, some chiral β-thio-substituted phosphines have been used as metal ligands in asymmetric catalysis and are listed in Sect. 5.3.

5.2.1
β-Sulfinyl Phosphonates and Phosphines

Optically active (*o*-alkylsulfinylaryl)phosphonates, as well as their analogues phosphines, have been recently prepared (Scheme 42). Following the [1,3]-

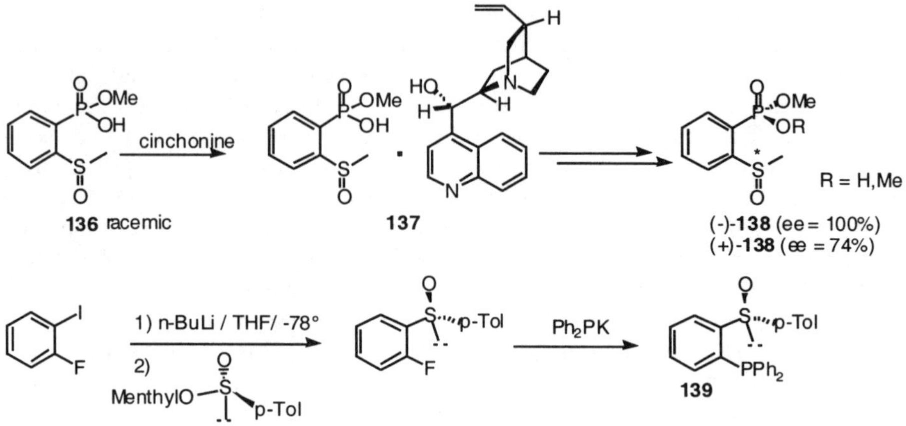

Scheme 42

sigmatropic rearrangement described in Sect. 3.3, non racemic (2-methylsulfinyl)phenylphosphonic acid mono- and dimethyl esters (–)- and (+)-**138** have been obtained by resolution of the corresponding racemic monoester **136** via its cinchonine diastereomeric salts **137** [85]. Besides, the enantiopure phosphine-sulfoxide ligand **139** has been obtained in two steps from 2-fluoroiodobenzene, menthyl *p*-tolyl sulfinate and potassium diphenylphosphide [86].

A sulfinyl-substituted phosphine chelate **140,** containing three carbons, one phosphorus and one sulfur stereogenic center, has been prepared by a Pd-promoted asymmetric Diels-Alder reaction (Scheme 43) [87a]. Hetero Diels-Alder reaction between the same Pd-complex of the phospha-substituted cyclic diene and a thiocarbonyl dienophile generates only one stereomeric cycloadduct [87b]. The same group describes the synthesis of both enantiomerically pure forms of the phosphine ligand (*R*)- and (*S*)-**141** by cycloaddition between 3,4-dimethyl-1-phenylphosphole and phenyl vinyl sulfide in the presence of a chiral organopalladium template [88a]. In vitro cytotoxicity tests of the gold(I) complexes (*S*)- and (*R*)-**142** of these phosphines have shown their potential as anticancer agents [88b].

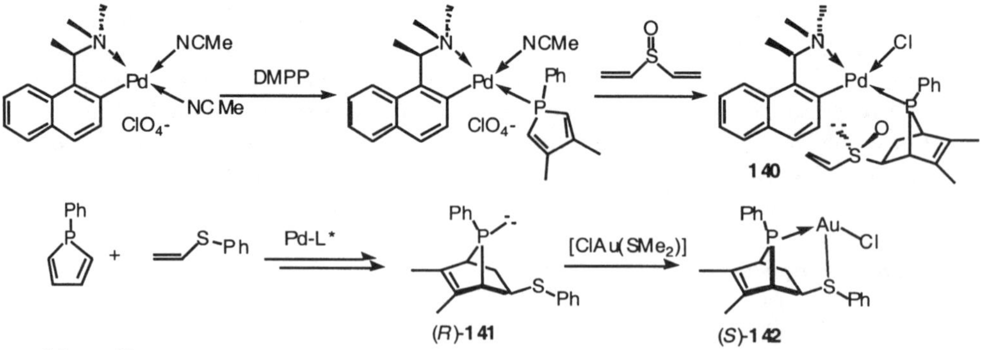

Scheme 43

5.2.2 Phosphono Substituted Chiral Thiazolidines

The 4-thiazolidinyl phosphonates **143** (Scheme 44) are known for their therapeutical properties, in particular as anti-inflammatory agents [5, 89]. Their asymmetric synthesis by hydrophosphonylation of 3-thiazolines has been described using various chiral auxiliaries: chiral phosphites such as (2*S*, 4*R*)-2H-2-oxo-5,5-dimethyl-4-phenyl-1,3,2-dioxaphosphorinane (de=2–8%) [90] or BINOL-phosphite (de=65–90%) [91] and also chiral catalyst such as titanium or lanthanide chiral complexes (ee=29–98%) [92]. Hydrophosphonylation of C2-chiral 3-thiazolines has also been performed (de=32–38%) [93].

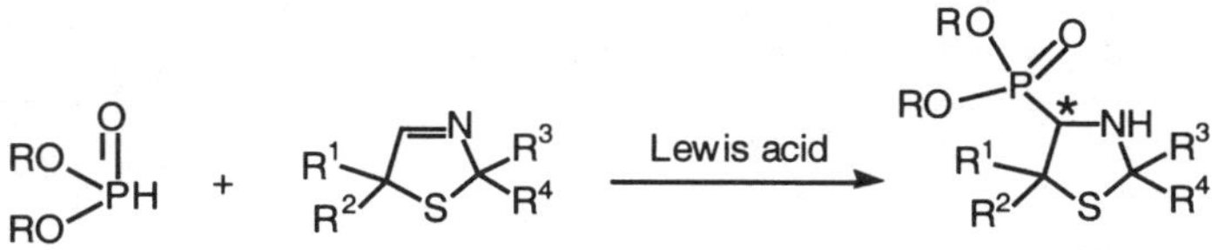

Scheme 44

5.3 Chiral P,S-Ligands in Asymmetric Catalysis

Many recent books and reviews cover the literature concerning asymmetric catalytic processes, often classified by type of reaction and including the preparation and the role of chiral ligands [94]. Among chiral bidentate mixed-donor ligands containing a phosphine function such as P,O-, P,N-, and P,S-ligands, the latter are less described because sulfur has long been considered as catalyst poison [95]. This idea has been partly denied by several recent examples of ligands containing phosphorus and sulfur as donor atoms for metal catalysts. A selection of the main structures synthesized for this purpose is presented in Scheme 45.

The synthesis of heterobidentate axially chiral P,S-binaphthyl derivatives **144** has been reported [96]. This class of ligands has been tested in the enantioselective Rh(I)-catalyzed hydroformylation (10–25% ee), hydrosilylation of styrene (54% ee) [97], addition of diethylzinc to aldehydes (14% ee), conjugate addition of Grignard reagents (20% ee) and Pd-catalyzed allylic substitution [98]. Starting from the enantiopure ferrocenylphosphine thiol **145a** and the diastereoisomers of *N*-methyl ephedrine, four corresponding sulfides **145b** have been isolated and used in the Au(I)-catalyzed synthesis of oxazolines (13–89% ee) [99]. An analogous ferrocenylphosphine ligand **146** derived from thioglucose has also been prepared and tested in the Pd-catalyzed allylic substitution (88% ee) [100]. Kagan et al. have synthesized and used 1,2-disubstituted ferrocene bearing both a chiral *p*-toluenesulfinyl group (obtained from menthyl *p*-toluenesulfinate) and a phosphinoborane group for the preparation of enantiopure ferrocenyl bis(phosphanes) ligands with planar chirality [101, 102]. Analogues bearing a chiral *tert*-butyl sulfinyl group (obtained by enantioselective oxidation of the corresponding sulfide) has also been prepared [103]. However, to our knowledge these

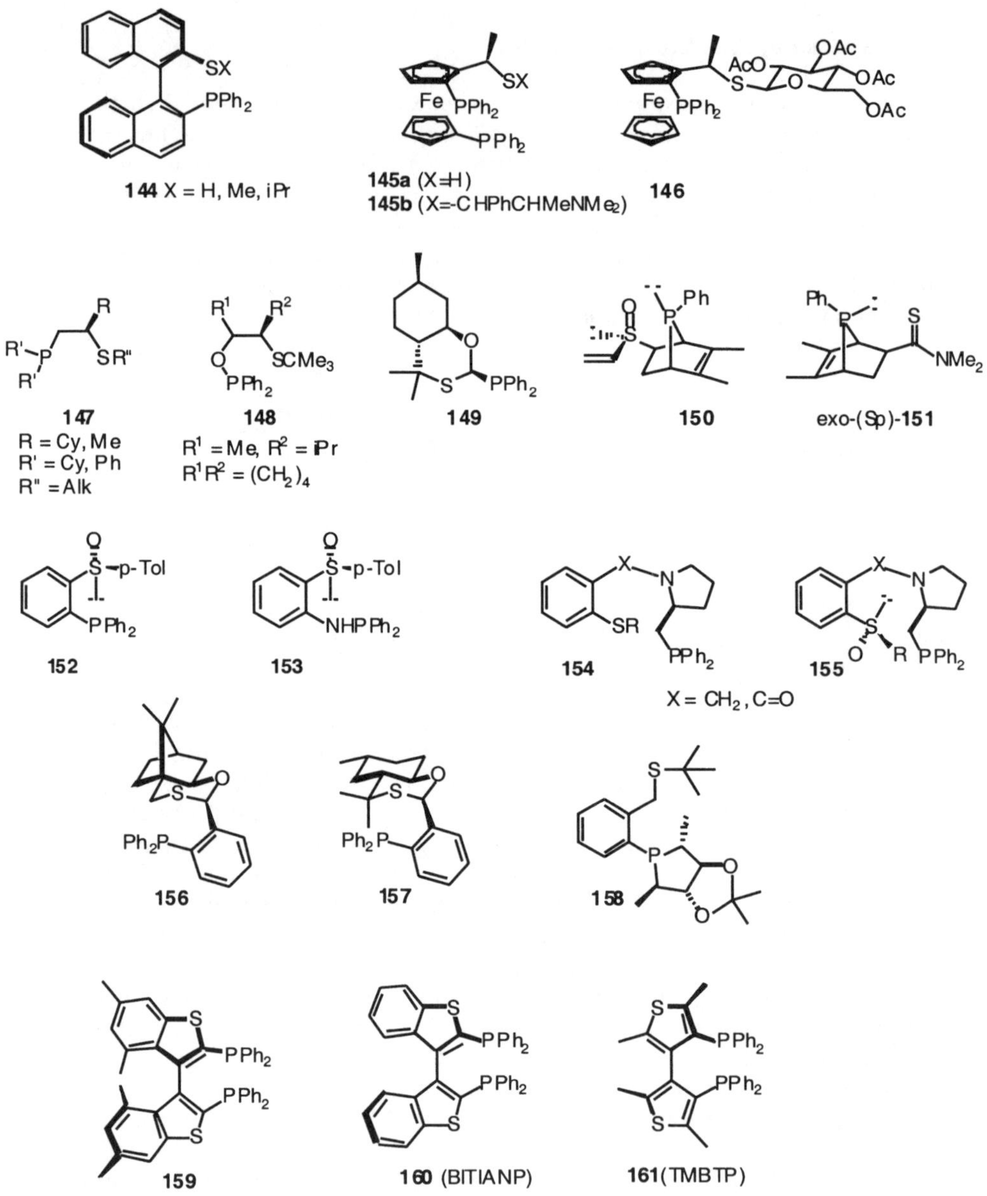

Scheme 45

phosphino-ferrocenyl sulfoxides have not been tested as ligands in asymmetric catalysis. New P,S-ligands **147**, formed by the opening of chiral episulfides with phosphorus nucleophiles, have been reported and their Rh-complexes have been used in the asymmetric hydrogenation of α-enamide methyl esters, providing enantioselectivities up to 51% [95a]. Phosphinite-sulfide bidentate ligands **148** described by Evans et al. give excellent results (91–99% ee) in Pd-catalyzed allylic substitution [104]. Starting from natural (+)-pulegone, the ligand **149** (PuPHOS) bearing the phosphino group and the sulfur on the same carbon has been synthesized and its dicobalt tetracarbonyl complex used in asymmetric intermolecular Pauson-Khand reaction [105]. The preparation, by Pd-promoted asymmetric Diels-Alder cycloaddition, of the optically active ligand **150** with five stereogenic centers and its chelation with Pd(II) has been studied [106]. Pd-complexes of P,S- heterobidentate phosphine ligands **151** *endo*-(*R*p) and *exo*-(*S*p) containing thiocarbonyl sulfur donor have also been described [107]. Enantiopure benzene sulfoxide ligands **152** and **153**, ortho-substituted by a phosphino or a phosphamino group, have been recently synthesized [86] and their rather good efficiency in Pd-catalyzed allylic substitution is demonstrated. The same authors have also described (*S*)-proline-derived ligands **154** and **155** with phosphanyl, sulfanyl, or sulfinyl chelating functionalities and enantiomeric excess up to 84% are observed for the Pd-catalyzed allylic alkylation [108]. For the same catalytic reaction, isoborneol- (or borneol) based phosphinooxathiane ligand **156** or pulegone-based ligand **157** have been successfully used (ee up to 90%) [109] whereas only 44% of enantiomeric excess is obtained with the monophospholane bearing pendant *t*-BuS group ligand **158** [110].

Although sulfur is unlikely to chelate the metal in this case, it is worth mentioning the axially chiral diphosphine ligands, based on *bi*-thienyl systems which increase the electronic density at phosphorus such as: **159** (used in Ru-catalyzed reduction of β-keto esters with 99% ee) [111a], BITIANP **160**, and TMBTP **161** (in a Pd-catalyzed Heck reaction, the regio- and enantioselectivity are high with **160** but low with **161**) [111b].

6
Conclusion

Although non-exhaustive, this review shows that a large variety of new difunctionalized mixed phosphorus and sulfur molecular structures have been described during the last ten years. It also demonstrates that such compounds are powerful synthetic tools or building blocks and, in some cases, molecules (or precursors of more complex molecules) with potential biological activities.

The association between the phosphonate and dithioester functions induces a very versatile reactivity. From the phosphorylated dithioesters, beside the thiophilic addition of organometallics and soft nucleophiles, which generates stabilized carbanions or ylides, the hetero Diels-Alder and thioacylation reactions give access to a variety of phosphorylated heterocycles (imidazolines, thiazolines, thiopyrans precursors of thiaglycosides, etc.). HWE reaction of phosphonodithioacetate derivatives (in particular phosphorylated thioamides derived from aminoacids or aminoalcohols) allows olefination of carbonyl compounds with functionalization. This chemistry is again enriched by introduction of fluorine on the carbon linking the P and S functions. In particular, the methyl monofluorophosphonodithioacetate is a promising building block for the synthesis of a variety of new monofluorophosphonates (including analogues of monophosphorylated nucleosides) and fluorovinylic derivatives.

Well known carbanionic sigmatropic rearrangements, applied to mixed P and S compounds, regio- and/or stereoselectively lead to new (α-sulfanylalkyl) or (β-sulfanylaryl) phosphonates, phosphine oxides, or phosphorodiamidates. In these difunctional compounds, chirality can be either introduced on the phosphorus, on the α-carbon, or on the sulfur atom.

It is worthy of note that α-sulfanyl phosphonic acids, which can now be obtained enantioselectively from corresponding α-hydroxyphosphonates, are analogues of the α-sulfanyl carboxylic acids, which, for some of them, are metallo-β-lactamases inhibitors [112]. To our knowledge, it does not seem that biological activities of α-sulfanyl phosphonic acids have been examined so far.

The (α-sulfanylalkyl)phosphonates and their difluorinated derivatives are also, by easy radical cleavage of the C–S bond, useful precursors of phosphonomethyl or phosphonodifluoromethyl radicals, which can be added to double bonds and so, introduced in a variety of structures. Besides, the use of phosphonodithioformates as radical trapping agents and their use as RAFT reagent for controlled polymerization open a new interesting field of application for these dithioesters.

In addition to the HWE reactions leading to asymmetric vinylic sulfoxides (not reviewed here), new recent applications of enantiopure α-sulfinylphosphonates, such as their use as precursors of optically active di-alkyl or arylalkyl sulfoxides and also of 1-sulfinyl-phosphonocyclopropanes, increases again their synthetic potential.

The last ten years have also seen β- or γ-thio-substituted phosphines (the sulfur function being a thiol, sulfide, or sulfoxide) increase in importance relative to their oxygen and nitrogen analogues as chiral bidentate ligands in asymmetric catalysis, and probably this is only a beginning. In particular, the use of the corresponding thio-substituted phosphonates, phosphine oxides, or phosphonodiamidates with chirality on phosphorus or/and sulfur (oxidizable into the sulfoxide function which can be either hard or soft ligand) would need further investigation. As far as the metal complexing properties are concerned, the elaboration of new metal sequestering drugs or radiopharmaceutics (technetium, rhenium complexes) is another potential field of applications for the variety of available polydentate phosphorus and sulfur ligands. Therefore, we are confident that the following years will again see new developments of the rich chemistry of mixed phosphorus and sulfur compounds.

7 References

1. (a) Brelière JC, Edmondts-Alts X, Garcia G (1983) Eur Patent 100 718; (1984) Chem Abstr 100:192078j; (b) Cristau HJ, Brahic C, Pirat JL (2001) Tetrahedron 57:9149
2. (a) Ash DE, Farrington GK, Ewaskiewicz JE, Shames SL, Wedler FC (1987) Fed Proc 46:2070; (b) Harde C, Neff KH, Nordhoff E, Gerbling KP, Laber B, Pohlenz HD (1994) Bioorg Med Chem Lett 4:273
3. Stanton JL, Ksander GM, de Jesus R, Sperbeck M (1993) Bioorg Med Chem Lett 4:539
4. (a) Magnin DR, Biller SA, Dickson JK, Lawrence M, Sulsky RB (1994) Eur Patent 595635; (b) Magnin DR, Biller SA, Chen Y, Dickson JK Jr, Fryszman OM, Lawrance RM, Logan

JVH, Sieber-McMaster ES, Sulsky RB, Traeger SC, Hsieh DC, Lan SJ, Rinehart JK, Harrity TW, Jolibois KG, Kunselman LK, Rich LC, Slusarchyk DA, Ciosek CP Jr (1996) J Med Chem 39:657
5. Andrews KJM (1981) Eur Patent 33919 Hoffmann-La-Roche; (1982) Chem Abstr 96:52498r
6. (a) Masson S (1995) Phosphorus Sulfur Silicon Relat Elem 95/96:127; (b) Masson S (1995) Rev Heteroatom Chem 12:69
7. (a) Mikolajczyk M, Balczewski P (1994) Adv Sulfur Chem 1:41; (b) Mikolajczyk M (1989) Rev Heteroatom Chem 2:19; (c) Mikolajczyk M (1994) S Afr J Chem 47:103
8. (a) Metzner P (1992) Synthesis 1185; (b) Metzner P, Thuillier A (1994) Sulfur reagents in organic synthesis. Academic Press, London; (c) Metzner P (1999) In: Page PCB (ed) Topics in current chemistry. vol. 204. Organosulfur chemistry I. Springer, Berlin Heidelberg New York, p 127
9. Grisley DW (1960) J Org Chem 26:2544
10. Bulpin A, Masson S, Sene A (1989) Tetrahedron Lett 30:3415
11. Bulpin A, Masson S, Sene A (1990) Tetrahedron Lett 31:1151
12. Bulpin A, Masson S (1992) J Org Chem 57:4507
13. Le Roy-Gourvennec S, Lemarié M, Masson S (1997) Phosphorus Sulfur Silicon Relat Elem 123:247
14. Bulpin A, Le Roy-Gourvennec S, Masson S (1994) Phosphorus Sulfur Silicon Relat Elem 89:119
15. Makomo H, Masson S, Saquet M (1994) Tetrahedron 50:10,277
16. Makomo H, Masson S Saquet M (1993) Phosphorus Sulfur Silicon Relat Elem 80:31
17. Heuzé B, Gasparova R, Heras M, Masson S (2000) Tetrahedron Lett 41:7327
18. Heras M, Gulea M, Masson S (2001) Chem Commun 611
19. Hartke K, Kunze O, Hoederath W (1985) Synthesis 960
20. (a) Barnikow G, Strickmann GS (1968) Z Chem 33; (b) Marvel CS, De Radzitzky P, Bader JJ (1955) J Am Chem Soc 77:5997
21. Hartke K, Hoederath W (1983) Sulfur Lett 1:191
22. Le Roy-Gourvennec S, Masson S (1995) Synthesis 1393
23. Prévôté D, Le Roy-Gourvennec S, Caminade AM, Masson S, Majoral JP (1997) Synthesis 1199
24. Sandrinelli F, Le Roy-Gourvennec S, Masson S, Rollin P (1998) Tetrahedron Lett 39:2755
25. Leflemme N, Marchand P, Gulea M, Masson S (2000) Synthesis 1143
26. (a) Al Badri H, Collignon N, Maddaluno J, Masson S (2000) Tetrahedron 3909; (b) Al Badri H, Collignon N, Maddaluno J, Masson S (2000) Chem Commun 1191
27. (a) Pfund E, Lequeux T, Vazeux M, Masson S (2002) Tetrahedron Lett 43:2033 (b) Pfund E, Lequeux T, Masson S, Vazeux M (2002) Org Lett 4:843
28. Pfund E, Lequeux T, Masson S, Vazeux M (2002) 20th International Symposium on the Organic Chemistry of Sulfur, Flagstaff, Arizona, USA
29. (a) Brückner R (1991) In: Trost BM, Fleming I, Pattenden G (eds) Comprehensive organic synthesis. Pergamon Press, New York, vol 6, p 873; (b) Marko I (1991) In: Trost BM, Fleming I, Pattenden G (eds) Comprehensive organic synthesis. Pergamon Press, New York, vol 3, p 916; (c) Brückner R (1993) Kontackte (French edn), p 3; (1994) Kontackte (French edn), p 2
30. Gulea-Purcarescu M, About-Jaudet E, Collignon N (1994) J Organometal Chem 464(1): C14-C16
31. Gulea-Purcarescu M, About-Jaudet E, Collignon N (1995) Tetrahedron Lett 36:6635
32. Denmark SE, Miller PC (1995) Tetrahedron Lett 36:6631
33. Yokomatsu T, Yamagishi T, Shibuya S (1995) Synlett 1035
34. Gulea-Purcarescu M, About-Jaudet E, Collignon N, Saquet M, Masson S (1996) Tetrahedron 52:2075
35. Makomo H, Saquet M, Simeon F, Masson S, About-Jaudet E, Collignon N, Gulea-Purcarescu M (1996) Phosphorus Sulfur Silicon Relat Elem 109/110:445
36. Makomo H, Masson S, Putman D, Saquet M, Simeon F, About-Jaudet E, Collignon N (1996) Phosphorus Sulfur Silicon Relat Elem 112:193–202
37. Lemée L, Gulea-Purcarescu M, Masson S, Saquet M, Collignon N (1999) Heteroatom Chem 10:281
38. Makomo H, Masson M, Saquet M (1993) Tetrahedron Lett 34:7257

39. Gulea M, Marchand P, Masson S, Saquet M, Collignon N (1998) Synthesis 1635
40. Simonneaux G, Galardon E, Paul-Roth C, Gulea M, Masson S (2001) J Organomet Chem 617/618:360
41. Marchand P, Gulea M, Masson S, Saquet M, Collignon N (2000) Org Lett 2:3757
42. Moore JD, Sprott KT, Hanson PR (2001) Synlett 605–608
43. Masson S, Saquet M, Marchand P (1998) Tetrahedron 54:1523
44. Takeda K, Sumi K, Hagisawa S (2001) J Organomet Chem 611:449
45. (a) Melvin LS (1981) Tetrahedron Lett 22:3375; Dhawan B, Redmore D (1984) J Org Chem 49:4018; (b) Dhawan B, Redmore D (1985) Synth Commun 15:411; (c) Dhawan B, Redmore D (1989) Phosphorus Sulfur Silicon Relat Elem 42:177
46. Garbers HV, Modro TA (1990) Heteroatom Chem 1:241
47. (a) Masson S, Saint-Clair JF, Saquet M (1993) Synthesis 485; (b) Masson S, Saint-Clair JF, Saquet M (1994) Tetrahedron Lett 35:3083; (c) Masson S, Saint-Clair JF, Saquet M (1996) Bull Soc Chim Fr 133:951
48. (a) Mauger C (2000) PhD thesis, University of Caen; (b) Mauger C, Vazeux M, Albinati A, Kozelka J (2000) Inorg Chem 704
49. Watanabe M, Date M, Kawanishi K, Akiyoshi R, Furukawa S (1991) J Heterocyclic Chem 28:173
50. Legrand O, Brunel JM, Buono G (2000) Tetrahedron 56:595
51. Au-Yeung TL, Chan KY, Haynes K, William ID, Yeung LL (2001) Tetrahedron Lett 42:457
52. (a) Brunel JM, Constantieux T, Legrand O, Buono G (1998) Tetrahedron Lett 39:2961; (b) Legrand O, Brunel JM, Buono G (1998) Tetrahedron Lett 39:9419; (c) Legrand O, Brunel JM, Buono G (2000) Tetrahedron Lett 41:2105
53. Barton DHR, Géro SD, Quiclet-Sire B, Samadi M (1992) Tetrahedron 48:1627
54. Barton DHR, Géro SD, Holliday P, Quiclet-Sire B (1996) Tetrahedron 52:8233
55. Wnuk SF, Robins MJ (1996) J Am Chem Soc 118:2519
56. Balczewski P (1995) Phosphorus Sulfur Silicon Relat Elem 104:113
57. Balczewski P, Pietrzykowski WM, Mikolajczyk M (1995) Tetrahedron 51:7727
58. Balczewski P (1997) Tetrahedron 53:2199
59. Lequeux T, Lebouc F, Lopin C, Hongli Yang, Gouhier G, Piettre SR (2001) Org Lett 3:185
60. Levillain J, Masson S, Hudson A, Alberti A (1993) J Am Chem Soc 115:8444
61. Alberti A, Benaglia M, Della Bona MA, Macciantelli D, Heuzé B, Masson S, Hudson A (1996) J Chem Soc Perkin Trans 2 1057
62. Alberti A, Benaglia M, Hapiot P, Hudson A, Le Coustumer G, Macciantelli D, Masson S (2000) J Chem Soc Perkin Trans 2 1908
63. (a) Alberti A, Benaglia M, Bonora M, Bozatta V, Hudson A, Macciantelli D, Masson S (1998) Polym Degrad Stab 62:559; (b) Laus M, Papa R, Sparnacci K, Alberti A, Benaglia M, Macciantelli D (2001) Macromolecules 34:7269
64. Alberti A, Benaglia M, Hudson A, Macciantelli D, Masson S (2001) Res Chem Intermed 27:659
65. Marchand P, Gulea M, Masson S, Saquet M, Collignon N (2000) Org Lett 2:3757
66. Marchand P, Gulea M, Masson S, Averbuch-Pouchot MT (2001) Synthesis 1623
67. Kawashima T, Kojima S, Inamoto N (1994) Bull Chem Soc Jpn 67:2603
68. Denmark SE, Chen CT (1992) J Am Chem Soc 114:10,674
69. Lawrence RM, Biller SA, Dickson JK Jr, Logan JVH, Magnin DR, Sulsky RB, DiMarco JD, Gougoutas JZ, Beyer BD, Taylor SC, Lan SJ, Ciosek CP Jr, Harrity TW, Jolibois KG, Kunselman LK, Slusarchyk DA (1996) J Am Chem Soc 118:11,668
70. Brandi A, Cicchi S, Goti A, Pietrusiewicz KM (1991) Tetrahedron Lett 32:3265
71. Mikolajczyk M, Midura WH, Grzejszczak S, Zatorski A, Chefczynska A (1978) J Org Chem 43:473
72. Mikolajczyk M, Midura WH, Grzejszczak S, Montanari F, Cinquini M, Wieczorek MW, Karolak-Wojciechowska J (1994) Tetrahedron 50:8053
73. Mikolajczyk M, Zatorski A, Grzejszczak S, Costisella B, Midura WH (1978) J Org Chem 43:2518
74. Shibata N, Matsugi M, Kawano N, Fukui S, Fujimori C, Gotanda K, Murata K, Kita Y (1997) Tetrahedron: Asymmetry 8:303

75. Mikolajczyk M, Midura WH, Wladislaw B, Biaggio FC, Marzorati L, Wieczorek MW, Blaszczyk J (1997) Tetrahedron 53:2959
76. Van Steenis JH, Boer PWS, Van der Hoeven HA, Van der Gen A (2001) Eur J Org Chem 911
77. Adrio J, Carretero JC (1999) J Am Chem Soc 121:7411
78. Cardellicchio C, Fracchiolla G, Naso F, Tortorella P (1999) Tetrahedron 55:525
79. (a) Aggarwal VK, Barrell JK, Worrall JM, Alexander R (1998) J Org Chem 63:7128; (b) Aggarwal VK, Barrell JK, Worrall JM, Alexander R (1999) Phosphorus Sulfur Silicon Relat Elem 153/154:337
80. (a) Mikolajczyk M, Midura WH (1992) Tetrahedron: Asymmetry 3:1515; (b) Midura WH, Mikolajczyk M (1994) Phosphorus Sulfur Silicon Relat Elem 95/96:397
81. Midura WH, Krysiak JA, Mikolajczyk M (1999) Tetrahedron 55:14,791
82. (a) Midura WH, Krysiak JA, Wieczorek MW, Majzner WR, Mikolajczyk M (1998) Chem Commun 1109; (b) Midura WH, Mikolajczyk M (2002) Tetrahedron Lett 43:3061
83. Patel RN, Banerjee A, Szarka LJ (1997) Tetrahedron: Asymmetry 8:1055
84. Gulea M, Hammerschmidt F, Marchand P, Masson S, Pisljagic V, Wuggenig F (2003) Tetrahedron: Asymmetry, in press
85. Mauger C, Masson S, Vazeux M, Saint-Clair JF, Midura WH, Drabowicz J, Mikolajczyk M (2001) Tetrahedron: Asymmetry 12:167
86. Hiroi K, Suzuki Y, Abe I, Kawagishi R (2000) Tetrahedron 56:4701
87. (a) Siah SY, Leung PH, Mok KF (1995) J Chem Soc Chem Commun 1747; (b) Qin Y, White AJP, Williams DJ, Leung PH (2002) Organometallics 21:171
88. (a) Leung PH, Loh SK, Mok KF, White AJP, Williams DJ (1996) J Chem Soc Dalton Trans 4443; (b) Song Y, Vittal JJ, Chan SH, Leung PH (1999) Tetrahedron: Asymmetry, 10:1433
89. Drauz K, Koban HG, Martens J, Schwarze W (1985) Liebigs Ann Chem 448
90. Gröger H, Martens J (1996) Synthetic Commun 26:1903
91. Schlemmminger I, Lützen A, Willecke A, Maison W, Koch R, Saak W, Martens J (2000) Tetrahedron Lett 41:7285
92. (a) Gröger H, Saida Y, Arai S, Martens J, Sasai H, Shibasaki M (1996) Tetrahedron Lett 37:9291; (b) Gröger H, Saida Y, Sasai H, Yamaguchi K, Martens J, Shibasaki M (1998) J Am Chem Soc 120:3089; (c) Schlemmminger I, Saida Y, Gröger H, Maison W, Durot N, Sasai H, Shibasaki M, Martens J (2000) J Org Chem 65:4818; (d) Yamakoshi K, Harwood SJ, Kanai M, Shibasaki M (1999) Tetrahedron Lett 40:2565
93. Gröger H, Martens J, Goerlich JR, Schmutzler R (1997) Phosphorus Sulfur Silicon Relat Elem 128:153
94. (a) Jacobsen EN, Pfaltz A, Yamamoto H (1999) Comprehensive asymmetric catalysis I-III. Springer, Berlin Heidelberg New York; (b) Noyori R (1994) Asymmetric catalysis in organic synthesis. Wiley, New York; (c) Ojima I (2000) Catalytic asymmetric synthesis, 2nd edn, VCH Weinheim
95. (a) Hauptman E, Fagan PJ, Marshall W (1999) Organometallics 15:2419; (b) McCarthy M, Guiry PJ (2001) Tetrahedron 57:3809 and references cited therein
96. Gladiali S, Dore A, Fabbri D (1994) Tetrahedron: Asymmetry 5:1143
97. Gladiali S, Pirri G, Pulacchini S, Medici S, Dore A, Fabbri D (1998) Conference Proceedeing 11th ISHC 268
98. Kang J, Yu SH, Kim JI, Cho HG (1995) Bull Korean Chem Soc 16:439
99. Togni A, Häusel R (1990) Synlett 10:633
100. Albinati A, Pregosin PS, Wick K (1996) Organometallics 15:2419
101. Riant O, Argouarch G, Guillaneux D, Samuel O, Kagan HB (1998) J Org Chem 63:3511
102. Argouarch G, Samuel O, Riant O, Daran J-C, Kagan HB (2000) Eur J Org Chem 2893
103. (a) Hua DH, Lagneau NM, Chen Y, Robben PM, Clapham G, Robinson PD (1996) J Org Chem 61:4508; (b) Lagneau NM, Chen Y, Robben PM, Sin HS, Takasu K, Chen JS, Robinson PD, Hua DH (1998) Tetrahedron 54:7301
104. Evans DA, Campos KR, Tedrow JS, Michael FE, Gagné MR (1999) J Org Chem 64:2994
105. Verdaguer X, Moyano A, Pericàs MA, Riera A, Maestro MA, Mahia J (2000) J Am Chem Soc 122:10,242
106. Siah SY, Leung PH, Mok KF, Drew MGB (1996) Tetrahedron: Asymmetry 7:357
107. Leung PH, Qin Y, He G, Mok KF, Vittal JJ (2001) J Chem Soc, Dalton Trans 309

108. Hiroi K, Suzuki Y, Abe I (1999) Tetrahedron: Asymmetry 10:1173
109. (a) Hiroto N, Yuko O, Hiroshi H (2000) Tetrahedron Lett 41:4615; (b) Hiroto N, Yuko O, Mariko Y, Hiroshi H (2001) J Org Chem 66:620
110. Yan YY, RajanBabu TV (2000) Org Lett 2:199
111. (a) Benincori T, Brenna E, Sannicolo F, Trimarco L, Antognazza P, Cesarotti E, Demartin F, Pilati TJ (1996) J Org Chem 61:6244; (b) Tietze LF, Thede K, Sannicolo F (1999) Chem Comm 1811. See also for this family of ligands: Benincori T, Cesarotti E, Piccolo O, Sannicolo F (2000) J Org Chem 65:2043 and references (9) cited therein
112. Kumar I, Jolly RS (2001) Org. Lett 3:283 and references cited therein

Author Index Volume 201–229

Author Index Vols. 26–50 see Vol. 50
Author Index Vols. 51–100 see Vol. 100
Author Index Vols. 101–150 see Vol. 150
Author Index Vols. 151–200 see Vol. 200

The volume numbers are printed in italics

Subject Index

Printing: Saladruck Berlin
Binding: Stürtz AG, Würzburg